THE UNIVERSITY OF
WINCHESTER

Trauma in Contemporary Literature

"This collection of essays, beginning with Cathy Caruth's poetic, lyrical, and beautifully crafted reading of the metaphorics of ashes, moves through literatures of the traumas of apartheid South Africa, colonial violence, the wars in Vietnam and Iraq, the Holocaust, and 9/11 to analyze the global nature of the worst experiences of our times."

—*Brett Ashley Kaplan, University of Illinois, USA*

Trauma in Contemporary Literature analyzes contemporary narrative texts in English in the light of trauma theory, including essays by scholars of different countries who approach trauma from a variety of perspectives. The book analyzes and applies the most relevant concepts and themes discussed in trauma theory, such as the relationship between individual and collective trauma, historical trauma, absence vs. loss, the roles of perpetrator and victim, dissociation, *nachträglichkeit*, transgenerational trauma, the process of acting out and working through, introjection and incorporation, mourning and melancholia, the phantom and the crypt, postmemory and multi-directional memory, shame and the affects, and the power of resilience to overcome trauma. Significantly, the essays not only focus on the phenomenon of trauma and its diverse manifestations but, above all, consider the elements that challenge the aporias of trauma, the traps of stasis and repetition, in order to reach beyond the confines of the traumatic condition and explore the possibilities of survival, healing and recovery.

Marita Nadal is Professor of American Literature at the University of Zaragoza.

Mónica Calvo is Lecturer in American Literature at the University of Zaragoza.

Routledge Interdisciplinary Perspectives on Literature

Trauma in Contemporary Literature

Narrative and Representation

Edited by Marita Nadal and Mónica Calvo

Routledge
Taylor & Francis Group
NEW YORK   LONDON

First published 2014
by Routledge
711 Third Avenue, New York, NY 10017

and by Routledge
2 Park Square, Milton Park, Abingdon, Oxon OX14 4RN

*Routledge is an imprint of the Taylor & Francis Group,
an informa business*

Library of Congress Cataloging-in-Publication Data
   Trauma in contemporary literature : narrative and representation / edited
by Marita Nadal and Mónica Calvo.
       pages cm. — (Routledge interdisciplinary perspectives on literature ; 26)
    Includes bibliographical references and index.
   1. Psychology in literature.   2. Psychic trauma in
literature.   3. Wounds and injuries in literature.   I. Nadal, Marita, editor
of compilation.   II. Calvo, Mónica, editor of compilation.
    PN56.P93T74 2014
    809'.93353—dc23
    2013040126

ISBN13: 978-0-415-71587-4 (hbk)
ISBN13: 978-1-315-88050-1 (ebk)

Typeset in Sabon
by IBT Global.

Contents

PART III
Trauma and the Poblem of Representation

Acknowledgments

The editors would like to express their gratitude to all those who have participated in the preparation of this book, especially to the contributors and the institutions who have granted permission to print or reprint copyrighted material.

Thanks are also due to the project financed by the Spanish Ministry of Economy and Competitiveness (MINECO) (code FFI2012-32719) and by the Government of Aragón and the European Social Fund (ESF) (code H05), as their funds have helped pay for some of the necessary reprint permissions.

Trauma and Literary Representation
An Introduction*

Marita Nadal and Mónica Calvo

In "Beyond the Pleasure Principle" (1920), Freud defined traumatic neurosis as "a consequence of an extensive breach being made in the protective shield against stimuli" (1984, 303); in 1980, the American Psychiatric Association officially acknowledged the phenomenon of trauma, describing its effects as a new illness coined as "Post-Traumatic Stress Disorder" (PTSD). PTSD has been defined as the response to an event "outside the range of usual human experience" (1980, 236), which involves serious somatic and psycho-somatic disturbances. Although at first trauma was mainly associated with extremely unusual events, it has now become a powerful and complex paradigm that infiltrates contemporary history, literature, culture and critical theory. Thus, trauma theory developed in the 1990s in connection with the ethical turn that emerged in the previous decade and which affected literary theory and philosophy. In different manners, the increasing interest in trauma was a response to concerns about memory, politics, representation and ethics that became prominent at the turn of the twentieth century, and which have mainly focused on the extreme forms of violence and victimisation that came to light after World War II.

As many theorists have pointed out, "ours appears to be the age of trauma" (Miller and Tougaw 2001, 1), a "catastrophic age" (Caruth 1995, 11) characterised by the pervasiveness of collective and individual afflictions. Modernity is marked by the "sign of the wound," and "the modern subject has become inseparable from the categories of shock and trauma" (Seltzer quoted in Luckhurst 2008, 20). Trauma, an "all-inclusive" phenomenon (Caruth 1995, 4), also constitutes a fascinating cultural paradigm because, as Roger Luckhurst notes, "it has been turned into a repertoire of compelling stories about the enigmas of identity, memory and selfhood that have saturated Western cultural life" (2008, 80).[1]

Trauma has become "a prevalent preoccupation in recent theory and criticism," even "an obsession," as Dominick LaCapra and other critics have pointed out (LaCapra 2001, x). Michael Rothberg, for instance, discusses the present concern with contemporary trauma in the light of the Holocaust: "exploring the recent fascination with the Holocaust means exploring a more general contemporary fascination with trauma, catastrophe, the fragility of memory, and the persistence of ethnic identity" (1996, 3). Although trauma can be taken as a key to approach contemporary experience, we should not conclude that we are all victims and everything is traumatic, as LaCapra has noted. Warning against the tendency to appropriate the victims' experience, he emphasises the importance of being "responsive to the traumatic experience of others"—of generating what he calls "empathic unsettlement" (2001, 41)—which, among other things, suggests a close link between trauma and ethics. However, even if this kind of appropriation is avoided, it is trauma that crosses limits, disrupts boundaries, and "threatens to collapse distinctions": "no genre or discipline 'owns' trauma as a problem or can provide definitive boundaries for it" (LaCapra 2001, 96).

This characteristic of excess produces disorientation but also fascination: trauma escapes understanding and resists representation, evoking a "radical intensity" that Michael S. Roth relates to dystopia and LaCapra to the sublime:

> the concept of trauma has come to perform some of the same functions that negative utopia or dystopia once did. Trauma, like utopia, designates phenomena that cannot be properly represented, but one characterized by radical intensity. A widespread longing for intensity has come to magnetize the concept of trauma, giving it a cultural currency far beyond the borders of psychology and psychoanalysis. Trauma has become the dystopia of the spirit, showing us much about our own preoccupations with catastrophe, memory, and the grave difficulties we seem to have in negotiating between the internal and external worlds. (Roth 2012, 90–91)

> In the sublime, the excess of trauma becomes an uncanny source of elation or ecstasy. (LaCapra 2001, 23)

LaCapra's warning against the tendency "to convert trauma into the occasion for sublimity" (2001, 23) brings to mind the points in common between both notions, which have given rise to a variety of critical and cultural works. Thus, both trauma and the sublime imply incommensurability, disruption, terror and pain; they mark the limits of reason and resist representation, forcing critical thought to a crisis.[2] LaCapra refers to Lyotard's emphasis on the un(re)presentability of the sublime (2001, 190), a subject that the latter explores in *The Differend* (1988) with regard to the trauma of the Holocaust: "In the differend, something 'asks' to be put into phrases and suffers from the wrong of not being able to be put into phrases right away" (Lyotard1988, 13). In terms of the sublime, the pain of the Holocaust is such that it exceeds our ability to express it. The notion of the apocalypse is another case in point: a source of awe and fascination, it can be related both to trauma and the Gothic, and also to the numinous and the sublime, as Avril Horner points out in the first section of this volume: apocalyptic desire may be a longing for the end, but also for change and transformation (Berger 1999, 34).

In *Sublime Historical Experience* (2005), Frank Ankersmit discusses historical consciousness, emphasising that experience comes before language: he is concerned with the representation of the past and with the ways to capture the immediacy of experience, which language distorts and, paradoxically, appears to block. In short, Ankersmit wants to discover a way to express the unrepresentable. Interestingly, he finds that immediacy of experience—beyond the borders of language—in the realm of trauma and the sublime, which he describes as follows: "In sum, trauma can be seen as the psychological counterpart of the sublime, and the sublime can be seen as the philosophical counterpart of trauma" (2005, 338). Significantly, the connection between trauma and the sublime foregrounds the problem of representation and the limits of language, and is also related to questions of time. Just as the sublime transcends time, linking past, present and future, trauma disrupts the mind's experience of time, so the subject endlessly wavers between the present and "a primary experience that can never be captured" (Luckhurst 2008, 5).

Paradoxically, the traumatic event is "fully evident only in connection with another place, and in another time," Cathy Caruth argues (1995, 8); "the traumatic event *is* its future" (see Caruth's chapter included in this book). Freud's concept of *Nachträglichkeit* (translated as "deferred action," "belatedness," "afterwardsness") encapsulates the paradoxical temporality of trauma, which implies a recurrent tension between the traumatic impact and its delayed response. As Jean Laplanche puts it:

> Trauma consists of two moments: the trauma, in order to be psychic trauma, doesn't occur in just one moment. First, there is the implantation of something coming from outside. And this experience, or the memory of it, must be reinvested in a second moment, and then it becomes traumatic. It is not the first act which is traumatic, it is the internal reviviscence of this memory that becomes traumatic. (Quoted in Caruth 2001, 1)

Thus, the peculiar temporal structure of trauma involves unfinishedness and repetition. Since the survivor experiences trauma *"one moment too late,"* s/he is forced to confront the primary shock over and over again, as Caruth notes in her analysis of Freud's "Beyond the Pleasure Principle" (1996, 62). In turn, this repetition compulsion fixates the patient to her/his trauma (Freud 1984, 282) and induces a ghostly relationship with the past that results in hauntedness, "stasis and entrapment" (Lifton 1980, 124). Dori Laub offers an accurate formulation of this problem:

> While the trauma uncannily returns in actual life, its reality continues to elude the subject who lives in its grip and unwittingly undergoes its

ceaseless repetitions and reenactments. The traumatic event, although real, took place outside the parameters of "normal" reality, such as causality, sequence, place and time. The trauma is thus an event that has no beginning, no ending, no before, no during and no after. This absence of categories that define it lends it a quality of "otherness", a salience, a timelessness and a ubiquity that puts it outside the range of associatively linked experiences, outside the range of comprehension, of recounting and of mastery. [. . .] The survivor, indeed, is not truly in touch either with the core of his traumatic reality or with the fatedness of its reenactments, and thereby remains trapped in both. (1992, 69)

As Dori Laub observes, the timelessness and ubiquity of trauma confer on it a spectral character that seems to preclude the way out, since the future is also trapped in its grip. In *Archive Fever* (1995), Derrida discusses the notion of the archive in the light of memory, history and Freud's work. In Derrida's approach, the archive, like psychoanalysis, attempts to record/ archive the past, but this recording involves the potential for its own elimination: "The archive always works, and *a priori*, against itself" (Derrida 1998, 12). As he argues, the concept of *Nachträglichkeit* is of crucial relevance, because it encompasses past, present and future and undermines the distinction among them:

Is it not true that the logic of the after-the-fact (*Nachträglichkeit*), which is not only at the heart of psychoanalysis, but even, literally, the sinews of all "deferred" (*nachträglich*) obedience, turns out to disrupt, disturb, entangle forever the reassuring distinction between the two terms of this alternative, as between the past and the future, that is to say, between the three actual presents, which would be the past present, the present present, and the future present? (Derrida 1998, 80)

In fact, Derrida's reading of Freud's *Nachträglichkeit* ties in with his previous contentions in *Specters of Marx* (1993), where he takes the spectre as the symbol for a critique of amnesiac modernity. For Derrida, the trope of the ghost "is perhaps the hidden figure of all figures" (1994, 150), because it throws time out of joint, producing a "radical untimeliness" that cannot be reversed. By its own nature, the ghost never dies: "it remains always to come and to come back" (Derrida 1994, 123). If, as Derrida argues, "the structure of the archive is *spectral*" (1998, 84), so is the phenomenon of trauma, with its ubiquity, enigmatic core and ceaseless returns.

As a failure of memory, trauma implies forgetting and repetition, as Freud notes in "Beyond the Pleasure Principle": since the traumatic event is not remembered, the patient has to *repeat* in the present (1984, 288). In Caruth's words, it is the literality of the traumatic event "and its *insistent return* which thus constitutes trauma and points towards its enigmatic core" (1995, 5; emphasis added). In "After the End" (included in the first section of this book), Caruth insists on the paradox inherent in the traumatic return:

> Trauma, and ultimately life and the drive itself, is an attempt to return that instead departs. This figure, this concept, this story—the story also, we recall, of the child who plays "fort/da" with his reel—is about memory and history, and it is also the concept archiving its own history, as it returns, and departs, from its origins.

It is precisely this desire to escape the pattern of repetition and "return to the authentic and singular origin" (Derrida 1998, 85) that fuels Derrida's *Archive Fever* and his analysis of Freudian psychoanalysis:

> [T]o be *en mal d'archive* [. . .] is to burn with a passion. It is never to rest, interminably, from searching for the archive right where it slips away. [. . .] It is to have a compulsive, repetitive, and nostalgic desire for the archive, an irrepressible desire to return to the origin, a homesickness, a nostalgia for the return to the most archaic place of absolute commencement. (Derrida 1998, 91)

Just as repetition is a central problem of trauma, so is its resistance to representation, as has been pointed out before with regard to the sublime. If, as Lacan argues, trauma is "a missed encounter with the Real" (1981, 50), and the Real resists symbolisation absolutely—"the Real doesn't stop not being written" (Lacan 1998, 59)—trauma constitutes the realm of the unspeakable and the unrepresentable. In that sense, the inaccessibility of trauma evokes "the always absent signified" of deconstruction and poststructuralist discourse, but with a difference, since, in contrast to the relativism of the latter, trauma emphasises its links with history. Ana Douglass and Thomas Vogler point out this relationship:

> the traumatic event [. . .] is that which cannot be anticipated or reproduced. It thus allows a return to the real without the discredited notions of transparent referentiality often found in traditional modes of historical discourse. This combination of the simultaneous undeniable reality of the traumatic event with its unapproachability offers the possibility for a seeming reconciliation between the undecidable text and the ontological status of the traumatic event as an absolute signified. (2003, 5)

In *The Return of the Real* (1996) Hal Foster discusses the relationship among representation, trauma and the Real as reflected in Andy Warhol's *Death in America*, taking Lacan's definition of the Real as a basis: significantly, the "traumatic realism" that Foster finds in Warhol's images is based on repetition. Foster explains this concept as follows:

As missed, the real cannot be represented; it can only be repeated, indeed it *must* be repeated. [. . .] repetition in Warhol is not reproduction in the sense of representation (of a referent) or simulation (of a pure image, a detached signifier). Rather, repetition serves to *screen* the real understood as traumatic. But this very need *points* to the real, and it is at this point that the real *ruptures* the screen of repetition. (1996, 132)

Paradoxically, even if the Real—and trauma—cannot be directly represented, there are strategies not straightforwardly referential that can prove appropriate to evoke the traumatic. Thus, Warhol's repetitive images, apparently depthless and devoid of affect, "not only reproduce traumatic effects: they also produce them" (Foster 1996, 132).

Michael Rothberg also employs the concept of "traumatic realism" to explore the "means and modes of representation" (2000, 2) with regard to the Holocaust, and, by extension, to trauma as a whole. In Rothberg's approach, traumatic realism is "a form of documentation and historical cognition attuned to the demands of extremity"; in it, "the claims of reference live on, but so does the traumatic extremity that disables realist representation as usual" (2000, 14, 106). Taking into account the problematics of representation and the relevance of realism, modernism and postmodernism as "persistent responses to the demands of history," Rothberg points out the importance of understanding these categories "as relational terms" rather than sequentially (2000, 10). In order to exemplify this interaction, Rothberg focuses on the notion of the constellation, which Walter Benjamin uses to describe "the in-between space that ties together the present and past." This notion, "a sort of montage" in which a variety of elements are put together through the act of writing, emphasises "the importance of *representation* in the interpretation of history" (2000, 10). Undoubtedly, the constellation proves to be a very appropriate concept for approaching the traumatic, since its conflation of diverse temporalities and repetition-*cum*-variation structure evoke both the untimeliness and the acting out inherent in trauma.

In *Multidirectional Memory* (2009), Rothberg tackles again the issues of trauma and memory from the perspective of polytemporality, in a way that echoes the notion of Benjamin's constellation: thus, the concept of multidirectional memory "is meant to draw attention to the dynamic transfers that take place between diverse places and times during the act of remembrance" and "acknowledges how remembrance both cuts across and binds together diverse spatial, temporal, and cultural sites" (2009, 11). Along similar lines, Steven Connor's concept of contemporality emphasises the "reciprocal filtering" between diverse temporalities, which makes it impossible to approach the present in isolation:

In contemporality, the thread of one duration is pulled constantly through the loop formed by another, one temporality is strained through another's mesh; but the resulting knot can itself be retied, and the filtered system also simultaneously refilters the system through which it is percolating. (1999, 31)

In "Not Now, Not Yet" (see the first section of this volume), Luckhurst takes Connor's and Rothberg's arguments as the point of departure for a discussion of the difficulties inherent in the representation of war. His contention that the contemporaneity of war must be approached through the refractions of the polytemporal—"one war time will always be seen through the lens of another"—suggests the ceaseless tension between diverse places and temporalities, and also the gap between the traumatic event and its narrative, which has to be accessed in an indirect, refracted way.

Thus, if the representation of the traumatic entails inadequacy or incompleteness—loss—it also runs the risk of trivialisation or betrayal, as several theorists have noted. Roth observes:

> Banalization of a trauma through narrative pleasure [. . .] threatens recovery, normality. [. . .] In so far as [. . .] representation is tied to narrative, the quality that makes an experience traumatic (that we cannot take it in through the mental schemes available to us) is lost in the telling. This loss can be felt as a cure; and it can be felt as a betrayal, a sacrilege, or a renewed act of violence. (2012, 83)[3]

Moreover, representation also implies transformation and, paradoxically, some degree of forgetting:

> there is no way to defeat the forgetting that comes with the recovery from trauma, and the most powerful and subtle forms of forgetting are narrative memory and history. Narrative memory, which is at the core of historical representation on paper or on film, *transforms* the past as a condition of retaining it. (Roth 2012, 85)

However, silence does not seem to be the appropriate alternative, since it entails repression and absolute forgetting. Despite "irredeemable losses" (Roth 2012, 85), and even if literature may suggest the ambivalence of the archive, "the literary (or even art in general) is a prime, if not the privileged, place for giving voice to trauma as well as symbolically exploring the role of excess" (LaCapra 2001, 190).[4] As Geoffrey Hartman has remarked: "Literary verbalization [. . .] still remains a basis for making the wound perceivable and the silence audible" (2003, 259). Furthermore, this theorist has pointed out the close connection between content and form that links literature and trauma, as Anne Whitehead notes. Thus, taking the two aspects of trauma as a basis—the traumatic event as content, and the symptomatic response to the event as form—Hartman concludes: "On the level of poetics, literal and figurative may correspond to these two types of cognition" (1995, 536). In a variety of manners, trauma fiction mimics the structure of trauma: it portrays the undecidability of the traumatic and, therefore, remains *suspended* "between its attempt to convey the literality of a specific event and its figurative evocation of the symptomatic response to trauma through formal and stylistic innovation" (Whitehead 2004, 162). Considering "the narrative/anti-narrative tension at the core of trauma," Luckhurst foregrounds the suspension quality of the literary by evoking Derrida's approach to literature: "[Derrida] argued that the peculiar specificity of literature was its ability to incorporate the languages of other discourses [. . .], yet suspend their strict protocols of meaning and reference for a time" (Luckhurst 2008, 80).

However, despite trauma's undecidability and literature's *"being-suspended"* (Derrida quoted in Luckhurst), it should not be assumed that "[t]o be in a frozen or suspended afterwards [. . .] is the only proper ethical response to trauma," as Luckhurst notes in the closing section of *The Trauma Question* (2008, 210). On the contrary, it is by working through and ethics that the path beyond trauma can be found.[5] If "the traumatic nature of history means that events are only historical to the extent that they implicate others", and "one's own trauma is tied up with the trauma of another," as Caruth argues, trauma may lead to an ethical encounter with the other, or echoing Lacan, to "an *ethical* relation to the real" (1996, 18, 8, 102). In *Trauma and Recovery*, psychiatrist Judith Herman concludes that "[r]esolution of the trauma is never final; recovery is never complete"; "trauma is redeemed only when it becomes the source of a survivor mission" (2001, 211, 207). Similarly, Dori Laub emphasises the central role of the survivors, who, within today's "culture of narcissism," are vehicles of an "inexorable historical transvaluation," the implications of which "we have yet to understand": "as *asserters of life out of the very disintegration and deflation of the old culture*, [the survivors] unwittingly embody a *cultural shock value* that has not yet been assimilated" (1992, 74). Somehow or other, belatedness—the failure to have awakened/reacted in time—can give rise to the ethical imperative to act and speak that awakens others.

It is in this light that the present volume gathers a wide spectrum of essays dealing with the literary representation of trauma and with the impulse, the need, to unravel the intricacies and undecidability of individual and collective traumas for the sake of both the victim's recovery and the ethical, empathic unsettlement of the other involved in the act of telling. Three distinct yet interconnected parts follow this introduction. Part I is wide in scope and lays the theoretical and contextual ground for an analysis of the narrative texts in parts II and III. It establishes the volume's point of departure—the view that we are living in a post-traumatic and post-apocalyptic age that lies in the ashes of history and continues to suffer from the wounds of the past and the globalised horrors of the present. Issues like the collective scope of individual trauma, put forth in Horner's essay, Caruth's Derridean approach to history and psychoanalysis or the use of refraction as a productive tool explored by Luckhurst are analysed in the subsequent parts. The chapters in Part II share their focus on the harmful effects of silence and on the need to come to terms with traumatic experiences by means of writing—what Suzette Henke (1998) denominates "scriptotherapy." The healing power of narrative is first tackled by Pellicer-Ortín, whose article on German born Jewish-British author Eva Figes interrogates the appropriateness of auto-biographical modes for working through the horror of surviving the Holocaust. Bayer's analysis of Jenny Diski's *Then Again* discusses the legacy of the Holocaust as undergoing a difficult phase as the memory is passed on to later generations, implying that insufficient debate about trauma can instigate mental suffering in subsequent generations. Bayer's attention to transgenerational memory and the pernicious effects of trauma are also the object of study in the essays that follow, based on postcolonial narratives. Herrero engages with Nicolas Abraham and Maria Torok's theory of the phantom—already introduced in Bayer's chapter—in her in-depth analysis of Zoë Wicomb's *Playing in the Light*. While showing the effects of transgenerational trauma and the damage done to post-apartheid children by their parents' shame and silence, Herrero sheds light on the power of literature as a gateway into the dark and convoluted recesses of the mind that enables the individual to articulate and negotiate socially unacknowledged traumas. In her study of Janette Turner Hospital's *The Last Magician* and *Oyster*, Fraile takes up the issue of unresolved mourning in the colonisation of Australia. Here, verbalising trauma constitutes the only possibility for resilience: the only way to avoid neurosis is for the individuals to modify their representations of traumatic experiences. This section is closed by Ibarrola-Armendáriz's study of Junot Díaz's *The Brief Wondrous Life of Oscar Wao* as a transgenerational immigrant family chronicle where the antihero's seriocomic story sheds light on the impact of collective traumas on diaspora communities—in this case, the "Trujillato" regime in the Dominican Republic (1930–61). Kenneth Thompson's theories on social psychology may help to interpret the direct line connecting past atrocities and present-day symptoms, and thereby grant some symbolic healing and reconstitution of the community.

While the previous part focuses on the power of writing as a path towards working through and healing, Part III explores the difficulties inherent in the process of narrativisation and telling as a productive means of moving beyond traditional approaches to trauma (and) fiction. Amfreville's opening essay presents Hilda Doolittle's *Tribute to Freud* as a diptych that mimics the structure of trauma by means of the chronological reshuffling of its stages—reminiscent of the all-informing process of *Nachträglichkeit*—and offers a deferred explanation to buried events, teaching us to re-read the past in the light of the present. In a similar vein, Kostova's points out the complex narrative strategies employed by the collective narrator in Eugenides' *The Virgin Suicides* as the novel oscillates between an impulse to deal with traumatic experiences and an anti-narrative delivery of such experiences. The next three chapters concentrate on the structure of repetition and variation—connected to the notions of fugue dynamics and Benjaminian constellations—as a means to represent trauma. Thus, Martínez-Alfaro analyses how the representation of trauma in Cynthia Ozick's "The Shawl" and "Rosa" is marked by a dynamics which can be likened to the fugue pattern of Paul Celan's "Todesfuge," while focusing on the important role that an empathic listener has for lessening the effects of trauma in the psychically wounded. Ganteau takes up Luckhurst's discussion of refraction in his reading of Julian Barnes' *A History of the World in 10¹/² Chapters* as a historical allegory in which the rewriting of the biblical Noah episode is used to re-inscribe the Shoah-related themes of selection and genocide, with the ethics of love as the only available asset to move towards working through. Onega argues that the self-reflexivity, thematic dispersion and structural fragmentation of J. M. Coetzee's *Dusklands* express the novel's ideological engagement and ethical responsibility in representing the unspeakable horrors of the Vietnam War and the colonisation of South Africa, but also of capitalism, colonial domination and environmental destruction. In Escudero-Alías' study of Sarah Waters' *The Night Watch*, the shame stemming from being a social and sexual outsider is examined as an insidious individual trauma through the re-elaboration of Abraham and Torok's notions of the phantom and introjection. Together with the analysis of alternative temporalities and reverse chronology in the novel, Silvan Tomkins' theory of affects is brought to the fore in order to reveal how the devastating effects of shame can be neutralised by resetting affects such as joy, excitement, interest and love. The volume closes with Arizti's analysis of the interdependence of private and public trauma in McEwan's *Saturday*. It develops Morson's ideas on narrative time and Semino's theory of possible narrative worlds in order to explore the paradoxical ways in which the novel both encourages and resists determinism in the face of traumatic events. While historical events are perceived through "backward causation," in the private sphere fictional temporality promotes responsibility, agency and freedom of choice.

A wish, and indeed a purpose, that is common to all the chapters in this volume, is that of opening the windows and letting gusts of fresh air blow through the halls of theoretical and analytical criticism. Whether hope lies

Figure I.1 Meltdown Morning, Neil Jenney (1975). Printed by permission of the Philadelphia Museum of Art.

in the content or in the frame is for the beholder to ascertain. Neil Jenney's *Meltdown Morning* (1975) perfectly metaphorises our standpoint: its black frame, which includes the painting's title in stencilled bold letters, is, Jenney insists, "as important as the painting" (quoted in Kramer 1981, 1), and provides a sense of narrative and time to the meticulously realistic pictorial image. The annihilation suggested by "Meltdown" gives the picture haunting overtones, calling attention to the background mushrooms of atomic explosion. Yet, not only is the catastrophe of nuclear destruction *mourned* by homophony with "morning" and hence worked through; it is the reference to morning, to the rebirth of dawn, that suggests new awakenings, hopes and the possibility of new life and light after destruction and trauma.

*The research carried out for the writing of this introduction is part of a project financed by the Spanish Ministry of Economy and Competitiveness (MINECO) (code FFI2012–32719). The authors are also grateful for the support of the Government of Aragón and the European Social Fund (ESF) (code H05).

NOTES

1. Many critics have emphasised the pervasiveness of trauma in the present age. In *The Trauma Question* (2008), Roger Luckhurst includes numerous references to their work. See also Paul Crosthwaite's *Trauma, Postmodernism, and the Aftermath of World War II* (2009), in which he contends that postmodernism has not neglected history, but has rather reformulated it in terms of trauma: thus, the postmodern is "convergent with the post-traumatic" (2009, 13).
2. See Edmund Burke's *A Philosophical Enquiry into the Origin of Our Ideas of the Sublime and Beautiful* (1757), where he conceptualises the relationship between terror and the sublime: "Whatever is fitted in any sort to excite the ideas of pain, and danger, that is to say, whatever is in any sort terrible, or is conversant about terrible objects, or operates in a manner analogous to terror, is a source of the *sublime*; that is, it is productive of the strongest emotion which the mind is capable of feeling. [. . .] Indeed terror is in all cases whatsoever, either more openly or latently the ruling principle of the sublime" (Burke 1968, 39, 58). See also Jean-Francois Lyotard's *Lessons on the Analytic of the Sublime* (1994) and *The Lyotard Reader* (1989): "As for a politics of the sublime, there is no such thing. It could only be terror" (1989, 204).
3. In this regard, see Caruth's analysis of the film *Hiroshima mon amour*, in which she addresses the problem of representation, emphasising the risk of betrayal ("*how not to betray the past*") and the indirectness of the film's historical portrayal (1996, 25–56).
4. See also Ulrich Baer's *Spectral Evidence: The Photography of Trauma* (2002), Jill Bennett's *Emphatic Vision: Affect, Trauma, and Contemporary Art* (2005), Michael S. Roth's "Why Photography Matters to the Theory of History," included in his volume *Memory, Trauma, and History* (2012), and Hal Foster's *The Return of the Real: The Avant-Garde at the End of the Century* (1996).

5. LaCapra refers to this "frozen or suspended afterwards"—resistance to working through—as "fidelity to trauma." In contrast, "[w]orking through trauma involves the effort to articulate or rearticulate affect and representation in a manner that may never transcend, but may to some viable extent counteract, a reenactment, or acting out, of that disabling dissociation" (2001, 22, 42).

WORKS CITED

American Psychiatric Association. 1980. *Diagnostic and Statistical Manual of Psychiatric Disorders*. Vol. 3 (DSM-III). Washington D.C.: American Psychiatric Association.

Ankersmit, Frank. 2005. *Sublime Historical Experience*. Stanford: Stanford University Press.

Baer, Ulrich. 2002. *Spectral Evidence: The Photography of Trauma*. Cambridge: The MIT Press.

Bennett, Jill. 2005. *Emphatic Vision: Affect, Trauma, and Contemporary Art*. Stanford: Stanford University Press.

Berger, James. 1999. *After the End: Representations of Post-Apocalypse*. Minneapolis: University of Minnesota Press.

Burke, Edmund. (1757) 1968. *A Philosophical Enquiry into the Origin of Our Ideas of the Sublime and Beautiful*. Edited by J. T. Boulton. Notre Dame, IN and London: University of Notre Dame Press.

Caruth, Cathy. 1995. "Introduction." In *Trauma: Explorations in Memory*. Edited by Cathy Caruth, 3–12. Baltimore, MD: The Johns Hopkins University Press.

———. 1996. *Unclaimed Experience: Trauma, Narrative, and History*. Baltimore, MD: The Johns Hopkins University Press.

———. 2001. "An Interview with Jean Laplanche." Accessed April 15 2013. http://pmc.iath.virginia.edu/text-only/issue.101/11.2caruth.txt.

Connor, Steven. 1999. "The Impossibility of the Present: Or, from the Contemporary to the Contemporal." *Literature and the Contemporary: Fictions and Theories of the Present*. Edited by Roger Luckhurst and Peter Marks, 15–35. Harlow and New York: Longman.

Crosthwaite, Paul. 2009. *Trauma, Postmodernism, and the Aftermath of World War II*. Houndmills, Basingtoke and New York: Palgrave MacMillan.

Derrida, Jacques. (1993) 1994. *Specters of Marx: The State of the Debt, the Work of Mourning and the New International*. Translated by Peggy Kamuf. New York: Routledge.

———. (1995) 1998. *Archive Fever: A Freudian Impression*. Translated by Eric Prenowitz. Chicago and London: University of Chicago Press.

Douglass, Ana, and Thomas A. Vogler, eds. 2003. "Introduction." In *Witness and Memory: The Discourse of Trauma*, 1–54. London and New York: Routledge.

Foster, Hal. 1996. *The Return of the Real: The Avant-Garde at the End of the Century*. Cambridge: Cambridge University Press.

Freud, Sigmund. (1920) 1984. "Beyond the Pleasure Principle." In *On Metapsychology*, Penguin Freud Library XI, edited by Angela Richards and translated by J. Strachey, 269–338. London and New York: Penguin.

Hartman, Geoffrey. 1995. "On Traumatic Knowledge and Literary Studies." *New Literary History* 26 (3): 537–63.

———. 2003. "Trauma within the Limits of Literature." *European Journal of English Studies* VII (3): 257–74.

Henke, Suzette. 1998. *Shattered Subjects: Trauma and Testimony in Women's Life-Writing*. London: Macmillan.

Herman, Judith Lewis. (1992) 2001. *Trauma and Recovery*. London: Pandora.

Kramer, Hilton. 1981. "Art View: Neil Jenney—Elegance with a Political Twist; Berkeley, Calif." *The New York Times*, May 17. Accessed June 15 2013. http://www.nytimes.com/1981/05/17/arts/art-view-neil-jenney-elegance-with-a-political-twist-berkeley-calif.html?pagewanted=1.

Lacan, Jacques. (1968) 1981. "The Unconscious and Repetition". In *The Four Fundamental Concepts of Psychoanalysis*. Translated by Alan Sheridan, 17–64. New York: Norton.

———. 1998. *The Seminar XX: Encore, On Feminine Sexuality, the Limits of Love and Knowledge, 1972–73*. Edited by Jacques Alain Miller and translated by Bruce Fink. New York: Norton.

LaCapra, Dominick. 2001. *Writing History, Writing Trauma*. Baltimore, MD and London: Johns Hopkins University Press.

Laub, Dori. 1992. "Bearing Witness, or the Vicissitudes of Listening." In *Testimony: Crises of Witnessing in Literature, Psychoanalysis, and History*. Edited by Shoshana Felman and Dori Laub, 57–74. New York and London: Routledge.

Lifton, Robert Jay. 1980. "The Concept of the Survivor." In *Survivors, Victims, and Perpetrators: Essays on the Nazi Holocaust*. Edited by J. E. Dimsdale, 113–26. New York: Hemisphere.

Luckhurst, Roger. 2008. *The Trauma Question*. London and New York: Routledge.

Lyotard, Jean François. 1988. *The Differend: Phrases in Dispute*. Translated by Georges Van Den Abbeele. Manchester: Manchester University Press.

———. 1989. *The Lyotard Reader*. Edited by Andrew Benjamin. Oxford: Basil Blackwell.

———. 1994. *Lessons on the Analytic of the Sublime*. Translated by E. Rottenberg. Stanford: Stanford University Press.

Miller, Nancy K., and Jason Tougaw. 2002. "Introduction: Extremities." In *Extremities: Trauma, Testimony, and Community*. Edited by Nancy K. Miller and Jason Tougaw, 1–24. Urbana and Chicago: University of Illinois Press.

Roth, Michael S. 2012. *Memory, Trauma, and History: Essays on Living with the Past*. New York: Columbia University Press.

Rothberg, Michael. 2000. *Traumatic Realism: The Demands of Holocaust Representation*. Minneapolis: University of Minnesota Press.

———. 2009. *Multidirectional Memory: Remembering the Holocaust in the Age of Decolonization*. Stanford: Stanford University Press.

Whitehead, Anne. 2004. *Trauma Fiction*. Edinburgh: Edinburgh University Press.

Part I

Global Trauma and the End of History

1 After the End

Psychoanalysis in the Ashes of History*

Cathy Caruth

> Memories of her village peopled by the dead turned slowly to ash and in their place a single image arose. Fire. (Toni Morrison, *A Mercy*)

In an essay of 1907, "Delusion and Dream in Wilhelm Jensen's *Gradiva*," Sigmund Freud analyses the novella *Gradiva: A Pompeiian Fantasy* (Jensen 1927) as a story exemplifying the principles of psychoanalysis laid out in *The Interpretation of Dreams*. In Jensen's story, a young archaeologist becomes obsessed with the figure of a walking woman on a bas-relief he has seen on a trip to Italy. He names her "Gradiva" and convinced by a dream that the woman died in Pompeii during the eruption of Vesuvius in A.D. 79, he travels to the ruined city in order to search for the singular traces of her toe-prints in the ash.

In this archaeological love story set amid the ruins of Pompeii, Freud finds an allegory for repression and the reemergence of repressed desire. In a later reading of Freud's text, the twentieth century philosopher Jacques Derrida, in his book *Archive Fever* (*Mal d'Archive*, 1995a), discovers, inside Freud's figure of the archaeological dig, what Derrida calls an "archival" drive, a pain and a suffering (*mal*) that bears witness to the suffering, and evil, of a unique twentieth-century history. Derrida proposes that the history of the twentieth century can best be thought through its relation to the "archive," a psychic as well as technical procedure of recording or of "writing" history that participates not only in its remembering but also in its forgetting.

At the heart of psychoanalysis, Derrida suggests, is the thinking of an archival drive that simultaneously yearns after memory and offers the potential for its radical elimination. Beginning from a reflection on the main argument of *Archive Fever* concerning the nature of the psychoanalytic archive, I will argue that the texts of Freud and Derrida, read together, ultimately enable a rethinking of the very nature of history around the possibility of its erasure. Moving beyond what Derrida explicitly suggests, I will also argue that these insights about history can ultimately be understood only from within the literary story of Norbert Hanold, the archaeologist, and in particular, the story of his dream. In what follows I will begin with Derrida's general reflections on the archive and ultimately turn to the story of the dream, which is perhaps also that of psychoanalytic dreaming more generally, to ask: What does it mean for history to be a history of ashes? And how does psychoanalysis bear witness to such a history?

A BURNING ARCHIVE

The problem of the archive as an immediate contemporary question of historical memory first emerges in *Archive Fever* in the opening "Insert," where Derrida links the archive to the urgency of twentieth-century history:

> Why reelaborate today a concept of the archive? In a single configuration, equally technical and political, ethical and juridical?
>
> This chapter will designate the horizon of this question discreetly, so burning is its evidence. The disasters that mark this end of the millennium are also archives of evil: dissimulated or destroyed, prohibited, diverted, "repressed." Their treatment is equally massive and refined in the course of civil or international wars, of private or secret manipulations. No one ever renounces—and this is the unconscious itself—the appropriation of power over the document, over its detention, retention, or its interpretation. (Derrida 1995b)[1]

The question of the archive is a question of "today"—of a particular historical period, a question with its own historical place—because it is linked, today, to the "disasters that mark the end of the millennium." These disasters are not simply the objects of archives, or objects that call out for archiving; they are also, themselves, unique events whose archives have been repressed or erased, and whose singularity, as events, can be defined by that erasure. They can indeed, themselves be called *"archives du mal"*—archives of evil (or suffering)—because they not only leave an impression, but hide their impression. They involve evil or suffering, that is, precisely because they hide or prohibit their own memory: because they are themselves "hidden or destroyed, prohibited, repressed." They consist precisely in hiding themselves; they become events insofar as they are, precisely, hidden.

The thinking of the archive is, in this sense, not only a thinking of memory but a thinking of history, and one that marks in particular, as I will argue, the historicity of the twentieth (and now twenty-first) centuries. This history is not, as one might traditionally expect, constituted by events that create their own remembrance, but by events that destroy their own remembrance. "Think of the debates around all the 'revisions," says Derrida, "and think of the seismic movements of historiography [. . .] of techniques in the constitution and treatment of so many 'dossiers'" (1995b, 2). This is why, I would suggest, psychoanalysis must be brought together with the thinking of the archive, because psychoanalysis has long been interested in the relation between history—personal and collective history—and the ways that its memory is suppressed or repressed: the ways that history is not available for immediate conscious access. Indeed psychoanalysis must itself be understood, Derrida argues, primarily *as* an archival science. Psychoanalysis, I would suggest, can thus help us to think, and perhaps witness, a new kind of event that is constituted, paradoxically, by the way it disappears.

Return and Repetition

The archival figure emerges from, and also reinterprets, Freud's own famous figure for psychoanalytic discovery, the metaphor of the archaeological dig.[2] From 1896 onward Freud had repeatedly represented the surprising encounter with the unconscious through an analogy in which the unconscious aspects of the mind are likened to a buried city that occasionally shows signs of its presence and eventually comes to light in analysis. But at the heart of the archaeological metaphor, Derrida notes, we often find a different kind of figure, not the figure of buried objects but rather of buried writing:

> [Psychoanalysis] does not, by accident, privilege the figures of the imprint and of imprinting. Installing itself often in the scene of the archaeological dig, its discourse concerns, first of all, the stock of "impressions" and the deciphering of inscriptions, but also their censorship and the repression, the repression and the reading of registrations. (1995b, 2)

If the archaeological project is the uncovering of an object, the archival task is the reading of an inscription. In this reading, psychoanalytic discourse does not only unveil a meaning of "impressions" and their "repressions" but also "installs *itself*" at the heart of the dig. The psychoanalyst's act of interpretation does not, therefore, simply reveal what has been repressed, but may also repress again what has been inscribed. "Is it not necessary," Derrida asks, "to distinguish the archive from that to which it is too often reduced, notably the experience of memory and the return to the origin, but also the archaic and the archeological, the memory of the dig, in other words, the search for lost time?" (1995b, 1–2).

Psychoanalysis does not permit a simple "return to the origin" because the impression it reads is not only left *for* psychoanalysis, but also left *on* psychoanalysis as it encounters the surprise of the inscription, and ultimately *by* psychoanalysis as it deciphers, as it leaves its own impressions, right at the site of the original ones, in a new archival act. The encounter with the archive is thus an act of interpretation that appears like a return, but is also an event that partially represses, as it passes on, the inscriptions it encounters; that passes on not only an impression but also, somewhat differently, its repression. The deciphering of desire, at the scene of the dig, is thus also the communication, and repression of desire, not necessarily of erotic desire, but rather of archival desire, "this fever, this presence, this desire [of Freud]," as Lacan described it (1998, 54). Through the act of its own archival drive, in other words, psychoanalysis reveals an "absolute desire for memory" (1995b, 3)—its own desire and perhaps *also* a desire at the heart of the erotic—that attempts to return to the past but to some extent always repeats and passes on, in its very act of interpretation, the ways in which the past has been erased.

As a thinking of the archive, psychoanalysis thus becomes witness to the strange notion of a *memory* that *erases,* a new notion of memory that, I would argue, is at the heart of the notion of "archive fever" (*mal d'archive*). Indeed, I would suggest, Derrida's description of the archive in psychoanalytic thought alludes to a very specific and historically situated archival discovery, Freud's encounter with "repetition compulsion" after WWI and his reformulation of the content and form of psychoanalytic theory around the notion of the "death drive," another term that "archive fever" arguably attempts to translate. Freud, as we recall, described, in *Beyond the Pleasure Principle*, his encounter with a kind of memory of events that erased, rather than produced conscious recall: the dreams and memories of the soldiers of WWI whose death encounters repeatedly returned to interrupt, rather than enter, consciousness. No longer capable of interpreting these memories as expressions of unconscious desire, Freud came to understand them as repetitions of the experiences that the soldiers could not grasp, a form of memory that, in enacting what it could not recall, also passed on a historical event that this memory erased. These memories, in other words, in repeating and erasing, did not *represent* but rather *enacted* history; they *made* history by also erasing it. They themselves were archival memories because they archived history by effacing it, and in effacing history, they also created it. The soldiers became, as it were, self-erasing inscriptions of history. Traumatic memory thus totters between remembrance and erasure, producing a history that is, in its very events, a kind of inscription of the past; but also a history constituted by the erasure of its traces.

Psychoanalysis can thus think the singularity of twentieth-century history, the new impression it makes, because, as an archival theory, it describes the way memory can *make* history precisely by *erasing* it. The notion of the archive, as I would thus interpret both Freud and Derrida, is a *change in modes of memory* that is also a *change in history*, a change that is "equally technological and political, ethical and juridical" (1995b, 1). It is this surprising change in memory and history, moreover, that is reflected in Freud's reconfiguration of the notion of the drive as a death drive. Although it is called a "drive," it is *also* a *new* discovery in history—a new shift in the nature of the historical archive—whose own past cannot be traced, because it is effaced, and is thus named only in terms of the way it erases the archive of memory, including the archive of its own memory:

This drive [. . .] always operates in silence, it never leaves an archive of its own. [I]ts silent vocation is to burn the archive and to incite amnesia [. . .]. (1995a, 12)

[E]nlisting the in-finite, archive fever touches on radical evil. (1995a, 19–20)

Between the shock of the memory that effaces, and the shock of the discovery of this memory, is the event of an erasure, and of a history, that carries the name of the death drive, which is also archive fever, because it is made up of memory and is about memory, it is about the burning desire for memory and the history of its burning up.

FREUD'S FEVER

How can psychoanalysis bear witness to this erasure, beyond repression, which is new to the twentieth century? We should note that the concept of archive fever, and its fundamental psychoanalytic ancestors, the notions of traumatic repetition and of death drive, themselves, as concepts, enact a kind of return and repetition, a memory and its erasure. Indeed, *Beyond the Pleasure Principle*, in its attempt to provide an economic understanding of the mind and in particular in its return to the notion of the memory trace, brings us back to Freud's first full attempt at a psychic system, the 1895 *Project for a Scientific Psychology*. Derrida, likewise, in *Archive Fever*, returns explicitly to his own earlier attempt, in his 1967 "Freud and the Scene of Writing," to read Freud's *Project*, and to link it to Freud's later work. It is on the level of the formation of psychoanalytic concepts, then, in the way that Freud's concepts inscribe a memory, archive their own history within themselves, that psychoanalytic discourse will bear witness to the history that psychoanalysis encounters. This is how, I believe, we can interpret Derrida's insight about the Freudian concept:

> The principle of the internal division of the Freudian gesture, and thus of the Freudian *concept of the archive*, is that at the moment when psychoanalysis formalizes the conditions of archive fever and of the archive itself, it repeats the very thing [. . .] which it makes its object. (1995a, 91)

The notion of the history *referred to* by the concept of "archive" is available only by studying the *history of the concept* of the archive, a story that occurs not on the level of a simple narration but, itself, requires a new temporal and historical modality.

What we find, indeed, when we return from *Beyond the Pleasure Principle* to its apparent origins in *The Project for a Scientific Psychology* is a concept of memory as inscription and repetition that cannot be located simply in the past of Freud's work. For the concept of the memory trace in the *Project* itself describes a form of memory that has no simple beginning, and which is a meeting between forces that is also a breaching and inscription, the marking of a path, and the deferral of quantity. Memory thus originates as its own deferral and also as its later repetition, a fundamental *deferral and repetition at the beginning*:

[T]he concepts of *Nachträglichkeit* and *Verspaetung*, concepts which govern the whole of Freud's thought and determine all his other concepts, are already present and named in the *Project*. The irreducibility of the "effect of deferral"—such, no doubt, is Freud's discovery (Derrida 1978, 203).[3]

The notion of the memory trace, in other words, "at the beginning" of Freud's itinerary, already anticipates the concept of repetition compulsion. Returning to the beginning, then, the concepts of *Beyond the Pleasure Principle* repeat, and erase, this past, erase it to the extent that *Nachträglichkeit* no longer operates as a term within this later text, precisely at the moment that repetition and delay nonetheless dominate its notion of human history.[4]

Indeed, the concept of *Nachträglichkeit* is the central concept, I would argue, that Derrida attempts to rename in the concept of the archive. This psychoanalytic concept of deferred action enacts its own deferred action and its own repetition throughout Freud's career, but in doing so it also both records and effaces its own past, and to a certain extent becomes erased from the psychoanalytic archive. We see this self-archiving and self-erasing act in *Beyond the Pleasure Principle*, where the notion of a deferred experience is newly figured as an "attempt to return" by consciousness that ultimately fails and departs into the repetitions of a future history. Trauma, and ultimately life and the drive itself, is an attempt to return that instead departs. This figure, this concept, this story—the story also, we recall, of the child who plays "fort/da" with his reel—is about memory and history, and it is also the concept archiving its own history, as it returns, and departs, from its origins.[5]

The psychoanalytic concept, I am trying to suggest, archives its own history and in so doing bears witness to the newness, and alterity, to the shock, of a history it cannot assimilate but only repeat. The twentieth-century history to which psychoanalysis bears witness through its own historical unfolding is the intersection of these two dimensions—of the concept that repeats, and of the memory that erases. Which is also to say that what Freud shows us, in the peculiar unfolding of his own discourse, is that the history of WWI can be understood, precisely, as a history constituted by the erasure of its own memory.

Disappearing History

In the structure and history of its concepts, psychoanalysis thus registers the impact of a new self-erasing history. To cite Robert Jay Lifton, psychoanalysis is, itself, a "survivor" of WWI.[6] But this survival, which gives psychoanalysis, we might say, its own "survivor mission," its mission to witness not only the individual but collective history—a collective history that is beginning to disappear even as it is being produced—is also structured by the principle of *Nachträglichkeit*, by the passage into the future that constitutes all traumatic experience. Indeed, the concept of return and departure, so central to *Beyond the Pleasure Principle*, will repeat itself, in *Moses and Monotheism*, now as the story of Moses attempting to return the Hebrews to Canaan. The Hebrews murder him, and depart, traumatically, into the future of Jewish monotheistic history.

At the center of *Archive Fever*, this story—which Derrida addresses through an encounter with the Jewish historian Yosef Hayim Yerushalmi's book *Freud's Moses: Psychoanalysis Terminable and Interminable* (1991)—will return, once again, as a story *about psychoanalysis*. What brings Derrida to this book by Yerushalmi is the question of whether or not psychoanalysis is a Jewish science, a question that, for Yerushalmi, means that it is a science of, and always opening up to, the future. But the return and departure that constitutes the survival of the Jews in *Moses and Monotheism*, read as a traumatic (archival) history, is also the story, I would argue, of psychoanalytic thought, which ultimately conceives both the erasure of the history it witnesses, and potentially, the erasure of its own history, that is, the very history of psychoanalysis *as witness*. This is why, I believe, Derrida ultimately turns to this work, and to Yerushalmi's book, to talk about the question of futurity that lies not only at the heart of all trauma—as deferral and future repetition, as an attempted return that instead departs—but also lies specifically at the heart of the archival history of psychoanalysis at this moment of its conceptualisation, in Vienna in the late thirties. If psychoanalysis, in its thinking of *Nachträglichkeit*, witnesses a history that is constituted by the erasure of memory, then Freud also, I would suggest, in *Moses and Monotheism*, conceives—or narrates—the possibility of a history constituted *by the erasure of its own witness*, a history that burns away the very possibility of conceiving memory, that leaves the future, itself, in ashes.

PSYCHOANALYSIS IN THE ASHES

Yet at this very point in *Archive Fever*, in the ashes of history, at its site, another story is told, the story of the German writer Wilhelm Jensen's *Gradiva: A Pompeiian Fantasy* and of Freud's reading of his essay in "Delusion and Dream in Wilhelm Jensen's *Gradiva*." We turn back, in Derrida's "Post-script" to this earlier Freud text, written in 1907 after *The Interpretation of Dreams*, to discover an encounter between Freud and a dream, or rather, between Freud and a literary text about dreams that returns us to the site of a disaster, and to the site of literature, *to the site of literature as archive*.

To recapitulate the story briefly, it is, as you will remember, the story of a young German archaeologist, Norbert Hanold, who becomes obsessed with the plaster cast of a marble Roman copy of a Greek bas-relief he has found in Italy and managed to purchase upon his return to Germany. Enthralled by the position of one of her feet, he names her "Gradiva" ("she who walks") and, having become convinced, in his fantasies, that she was from Pompeii, he decides, after a dream, that she was buried there during the eruption of A.D. 79. He ultimately returns to Pompeii to search for her traces in the ash, and, encountering a woman he believes is Gradiva's spectre, engages in a series of conversations with her that finally bring him to the realisation that she is in fact his old neighbor Zoe Bertgang, for whom his desire is now, finally and consciously, aroused.[7]

Freud, who had travelled to Naples in 1902, thinks of Gradiva, when he thinks of his new science, after encountering in the story, by surprise, his own discovery:

> There is [. . .] no better analogy for repression, by which something in the mind is at once made inaccessible and preserved, than burial of the sort to which Pompeii fell a victim and from which it could emerge once more through the work of spades. Thus it was that the young archaeologist was obliged in his phantasy to transport to Pompeii the original of the relief which reminded him of the object of his forgotten youthful love. The author was well justified, indeed, in lingering over the valuable similarity which his delicate sense had traced out between a particular mental process in the individual and an isolated historical event in the history of mankind. (Hertz 1997)[8]

The analogy of burial and preservation by which Freud had in 1896 characterised his own discovery of the unconscious, is rediscovered here in a literary text in a peculiarly self-reflexive way: first of all, as a figure used unconsciously and symptomatically by the main character, the archaeologist Norbert Hanold, a figure that represents his own unconscious processes. And second, as a figure used by the author of the story, who "traces out" the similarity between the burial of Pompeii and the "process of the mind." Character and author are thus both tracers of footprints in the ash of a catastrophe; they "linger" [*verweilen*] at this moment of discovery, a pleasurable erotic lingering and also one tinged with a Faustian threat.[9] Freud too, we could say, as a lover—not of a woman but of a science— confronted by Jensen's text, is "reminded" of "the forgotten object of his youthful love," psychoanalysis, or a "piece" of psychoanalysis, a piece of its "historical" if not material truth, which, "by analogy," he must not only make emerge once more "through the work of spades" (that is, dig up, with pleasure, as an archaeologist) but must also trace out, pursue with fright, *in the ashes.*

Freud's encounter with this text, therefore, does not involve an act of recognition, but rather a repetition that burns with a passion for discovery, *beyond* that of the character Norbert Hanold. Yet Hanold, himself is, already, no longer an archaeologist, but rather an archivist. "Hanold burns with archive fever," Derrida writes, "he has exhausted the science of archaeology [. . .] This science itself was of the past." Hanold returns to Pompeii not in order to return to life—the life of Gradiva, and ultimately the life of "Zoe"—but rather to find Gradiva's singular "traces," something he comes to understand, in the story at a moment of memory. At this moment in the story, Hanold stands alone in the silent streets of Pompeii, which suddenly seems peopled by the dead:

> [Derrida writes] Hanold understands everything. He understands why he had traveled through Rome and Naples. He begins to *know* what he did not then know, namely his "intimate drive" or "impulse." And this knowledge, this comprehension, this deciphering of the interior desire to decipher which drove him on to Pompeii, all of this comes back to him in an act of memory. He recalls that he came to see if he could find her traces, the traces of Gradiva's footsteps. (1995a, 98)

Hanold does not wish to see Gradiva, but to revisit what makes her memory possible: the traces of her steps. Remembering traces, he thus, I would suggest, also understands (or theorises), belatedly; like the Freud of both the *Project* and *Beyond the Pleasure Principle*, he attempts to return to the origin of his own memory as the origin of Gradiva's traces. As I would interpret this moment, staged by the philosopher as a moment of conceptualisation—of theorisation that echoes Freud's own belated creation of psychoanalytic theory—Hanold thus wishes to return to the *possibility* of Gradiva's memory, to what will guarantee that she is memorable *in the future*. That she can no longer be forgotten, erased, excised from the archive, that her impression will remain, and her history will intertwine itself with his.

In Freud's own act of encounter with Hanold's trace, I would also argue, in repeating Hanold's gesture, he feverishly repeats or "outbids" Hanold, for he too attempts to return, but does so in Hanold's (and his own) future: he attempts to return to the possibility of another memory, that of his own love of his youth, psychoanalysis, as the very possibility of testimony. Derrida writes of the moment, in Jensen's story, of feverish conceptualisation: "He dreams this irreplaceable place, the very ash, where the singular imprint, like a signature, barely distinguishes itself from the impression. And this is the condition of singularity, the idiom, the secret, testimony" (1995a, 99).

Hanold's dreaming, and Freud's after him, I believe, raise urgent questions, questions beyond their concepts, beyond the fever of conceptualisation, at this burning site: How can I bear witness, Hanold wonders, how can I guarantee the memory of Gradiva? How can I bear witness, how testify, Freud unconsciously thus asks, to the traces of psychoanalysis, that is, to the very possibility of finding traces, to the very possibility of memory, in the ash of this conceptualisation and in the ash of this memory?

Burning Dreams

Hanold, at this site, "dreams ," "he dreams [. . .] the very ash." What does it mean to dream? If we look at the larger context of the passage to which Derrida refers above, we notice that the object of Hanold's memory, of his reason for coming to Pompeii is, in fact, a dream. Even while Derrida writes that Hanold "dreams," the actual dream of Hanold, to which he alludes, is to a certain extent effaced from the philosopher's text, as the dream is also effaced from many critical texts on Jensen and Freud. It is, then, a dream that repeatedly recedes into unconsciousness, though it is, I would argue, at the very heart of Jensen's story. This is Hanold's first dream of Gradiva, while still in Germany, and the first moment that he actually sees her. I quote the entire context of the passage from Jensen in which Hanold, in Pompeii, comes to remember this dream. It begins as he stands in Pompeii's streets:

> Hanold looked before him down the Strada di Mercurio [. . .]
> Then suddenly—
> With open eyes he gazed along the street, yet it seemed to him as if he were doing it in a dream. A little to the right something suddenly stepped forth [. . .] and across the lava stepping-stones, which led from the house to the other side of the Strada di Mercurio, Gradiva stepped buoyantly.
> [. . .]
> As soon as he caught sight of her, Norbert's memory was clearly awakened to the fact that he had seen her here once already in a dream, walking thus, the night that she had lain down as if to sleep [. . .]. With this memory he became conscious, for the first time, of something else; he had, without himself knowing the motive in his heart [. . .] come to Pompeii to see if he could here find trace of her. (1927, 48–50)

Hanold's act of remembering the reason for his trip to Pompeii is part of a complex scene of dreaming and awakening. Hanold's remembering, his moment of conceptualisation, is a belated awakening to the fact that his seeing of her is a repetition, the repetition of a previous dream, which appears after the fact, that is, to be a dream of following traces.

   This "frightful" dream, indeed, appears to constitute the figurative centre of the story, and thus a kind of originary moment of Norbert's history. The dream in fact constitutes Norbert's first sighting of Gradiva, as well as the place in which he also first sees the destruction of Pompeii:

> In it he was in old Pompeii, and on the twenty-fourth of August of the year 79, which witnessed the eruption of Vesuvius. The heavens held the doomed city wrapped in a black mantle of smoke [. . .] [T]he pebbles and the rain of ashes fell down on Norbert also [. . .]. As he

stood thus [. . .] he suddenly saw Gradiva a short distance in front of him. [. . .] Violent fright forced from him a cry of warning. She heard it, too, for her head turned toward him so that her face now appeared for a moment in full view. [. . .] At the same time, her face became paler as if it were changing to white marble. [. . .] [H]astening quickly after her [. . .] he found his way to the place where she had disappeared from his view, and there she lay [. . .] as if for sleep, but no longer breathing [. . .]. [H]er features quickly became more indistinct as the wind drove to the place the rain of ashes. [. . .] When Norbert Hanold awoke, he still heard the confused cries of the Pompeiians who were seeking safety. (1927, 11–14)

Confronted with this "frightfully anxiety-producing dream," Freud suggests that it is an anxiety dream illustrative of his dream theory. The anxiety of the dream and the destruction of Pompeii illustrate the repressed, and returning erotic desire for Norbert's long-forgotten neighbor Zoe. Freud thus reads the dream through the archaeological metaphor, as both the burying and partial reappearance of desire, a wish-fulfilment.

Yet Freud's interpretation has an uncanny effect: for the dream is thus not only *read* archaeologically, but also becomes thereby a *staging* of its own formation, a staging of the burial of Pompeii which is the figure of repression. It is a dream of the origin of dreaming, of the possibility of knowing and not knowing, of figuring without consciousness, a mode of witnessing that originates in a catastrophe.

What kind of origin is this dream? Freud himself, at the end of his essay, famously adds an odd after-thought about another motive for the dream:

This was the wish, comprehensible to every archaeologist, to have been an eye-witness of that catastrophe of 79. What sacrifice would be too great, for an antiquarian, to realize this wish otherwise than through dreams! (1927, 255)

The wish to be an eye-witness of catastrophe is Freud's wish, here expressed on the level of manifest content in the dream. But Freud also, I would suggest, sees, and doesn't see, something in the witnessing that the dream narrates, an origin and a catastrophe that is not exactly a burial, and not simply of the past. Indeed, the figure of the destruction of Pompeii is not precisely, or not simply, a figure of burial, since the peculiarity of this "singular historical event" is that the destruction occurred not simply through burial but through burial *by ashes*, which is also a burning-up, a destruction that does not simply preserve but may also totally incinerate the bodies it buries.[10] At the origin of the figure of repression is the possibility of a complete erasure, which the archaeological analogy of burial and preservation—and the concept of repression that it shapes—itself erases and bypasses, passes over to pass on.

Indeed, as a narrative, the story of the dream tells not only of a burial but of Gradiva's walking, "stepping" and of Norbert Hanold's following of her steps. Step after step, the dream narrates a tracing of steps, which Hanold will remember, later, when his memory is "awoken," as the attempt to find the singular "traces of her toe-print in the ash."[11] The tracing is both Norbert Hanold's, then, and also "the author's," who had linked the "singular" destruction of Pompeii to the process of the mind, which is also Freud's "fine sense," as a theoretician whose concepts must also be figures, and often literary figures. Indeed, what Freud sees and doesn't see in this dream, what he attempts to return to but must repeat and erase, I would suggest, is the origin of his own theory, his own theory of dreams, not only in *The Interpretation of Dreams* but in the memory traces of the *Project*. These traces of the past—the past of Freud's own theory—are also traces of the future, insofar as the dream, a dream of burning that is also an awakening, points us toward the other kind of nightmares, the *nightmares that awaken* which are the traumatic dreams of *Beyond the Pleasure Principle*.

In these ashes, the figure that dominates is not so much what Freud points out but rather, as I see it, the figure of the trace, or even more the action of tracing, the step after step. These steps are dispersed throughout the story both in the dreams of thoughts of Norbert and on the level of the letter: in the *VorGANG* (mental process) that is also an *UnterGANG* (destruction) figured as the creation, and erasure, of the steps of Gradiva, who is also Zoe *BertGANG*, whose steps Norbert follows in the streets of Pompeii, after remembering his dream of ashes. But what, exactly, is the figure of a footstep in the ash? How can ashes sustain a print, when ashes are precisely that which may disperse and drift away? And what would it mean to leave a trace, or a remainder in that which is, itself, a remainder, the ash that is the burned up trace of what is incinerated? The figure of ash itself is, indeed, not only the substratum *for* a writing that has taken place, but the figure *of* a writing that is burning up.

On the site of these ashes, Freud writes a new kind of language. Faced by the dream, he is stirred to seek a saving figure in Zoe (whose name means "life," in Greek), but he gives us in fact another kind of figure, the imagination of an unimaginable erasure that is carried, I would argue, by all of his figures of deferral, of repetition, of return and departure, of *Nachträglichkeit,* of trauma. The language of trauma is the language of this absolute erasure, not imaginable in the past or present but always, as something missed, and about to return, a possibility, always, of a trauma in the future. This is what it means to say, I would suggest, that the traumatic event *is* its future, is its repetition as something that returns but also returns to erase its past, returns as something other than what one could ever recognise. A singular and new event, but in this case, in Freud's facing of the event in the dream, an event that undoes everything we have thought of as events, as history, because it is a future event that *undoes* its own future.

In his burning mission, I would ask, does Freud become witness to the ashes that surround him, or endure *as* the remainder that psychoanalysis may also become? Cinders are indeed "incubation of the fire lurking beneath the dust" (Derrida 1991, 59). To burn with archive fever: does it mean to bear witness, or to be ash?

Strange Witnesses

What, indeed, is the language, or figure, of cinders, which is the language of Freud and of Derrida, if it is "the annihilation of the capacity to bear witness"? (Derrida 2005). What of the readers, for example, who read the figure of ash? We are indeed, Derrida suggests in response to such a question by Elisabeth Weber, "strange witnesses [. . .] who do not know what they are witnessing. [. . .] witnesses to something they are not witness to" (Derrida and Weber 1995). I would suggest that there is something we could call the language of ashes, which is perhaps a new kind of language that is, for us today, marked historically, at least since the middle of the twentieth century.

Indeed, the figure of ash also refers us to events that may not have a simple referent, but are signs of the unimaginable past or the unimaginable future.[12] A future which, as the psychoanalyst Elaine Caruth once remarked, "hovers as a silence over every psychoanalytic session."[13] Which is why the figure of ashes, though without a simple referent, is without question marked by this twentieth-century historicity, this exposure that is now the condition of our history, and why it is in a literary text—a text that, *itself*, has no single referent, a text that can *figure* what it cannot *think*— that, in both the psychoanalyst's work and in the philosopher's work, these ashes, and this strange witness by Freud and Derrida, first emerge.

We could indeed translate *Archive Fever* in its "Postscript" into a kind of allegory, installing Derrida's French philosophical language at the heart of Freud's German psychoanalytic archive:

> By chance [*par chance*], I wrote these last words on the rim of Vesuvius, right near Pompeii, less than eight days ago. For more than twenty years, each time I've returned to Naples, I've thought of her [*Gradiva*]. (1995a, 97)

Derrida's reading of his own burning conceptualisation of the archive, born from this reading of Freud and Jensen, thus takes place from a thinking of chance, by a movement within philosophy toward the incalculable:

> Who better than Gradiva, I said to myself this time, the *Gradiva* of Jensen and of Freud, could illustrate [. . .] this concept of the archive, where it marks in its very structure [. . .] the formation of every concept, the very history of conception? (1995a, 97)

The origin of a history, the possibility of conceptualisation, of thought, which is also the condition of the possibility of memory, and thus of history—only happens to be thought by chance. It is this singular event of chance from which is born the chance of a writing, and of the writing of a new origination of thought, *after the end*.

But in inscribing his own writing on the rim of Vesuvius and at the heart of Freud's psychoanalytic archive, Derrida also (in part unwittingly) retells his own theoretical work *very like* the story of Hanold, who returns to Pompeii to seek after Gradiva. The philosopher thus writes his theory *inside* the literary story of Norbert Hanold, duplicating Hanold's return to Pompeii, which is where the actual sighting of Gradiva takes place, though his chance meeting of Gradiva, or of that "something" that first passes by his eyes as he stands in the burning streets of Pompeii precisely at midday. The thought of Gradiva is not only Derrida's thought *about* "the *Gradiva* of Jensen and of Freud" but also a Hanold-like thought *of* her. The chance takes place within the literary text and only insofar as Derrida, like Freud, encounters but also misses something crucial that he sees and does not see, the importance of the dream at the very heart of Jensen's story. In this way, Derrida's interpretation not only impresses its own trace upon Freud's and Jensen's texts, but is in its turn impressed upon *by* them—impressed into their service, we could say, as the site of witness where chance, thoughts and chance words arise. They arise as archive fever, but also as a literary fever beyond and below the concept, beyond and below the feverish attempt at conceptualisation.

Indeed, I would suggest that we can read this chance in Derrida's own words about thought and concept ("la conception même de la conception") insofar as they are inscribed in the place of Hanold's story, "right on the ashes" (à même la cendre), where "it no longer even makes sense to say 'the very ash' or 'right on the ash' (ou il n'y a même plus de sens à dire 'la cendre même' ou 'a même la cendre')" (1995a, 100). In Derrida's French, something repeats—the *même*—the very, the same—but it repeats differently each time, in each different context, by chance.

In order to understand this language, however, we must also recall that Hanold's story, the literary text of Wilhelm Jensen, is itself first told *within a dream*, the dream of Norbert Hanold, which is the dream of returning to Pompeii and seeing Gradiva, by chance, in the ashes. I submit that the philosopher, in turn, can be said to be dreaming throughout the entire work of *Archive Fever*, from its first pages: "I dream now," he says at the beginning of his book, "of having the time to submit for your discussion more than one thesis. [. . .] This time will never be given to me" (1995a, 5). Derrida's philosophic and literary dreaming returns, much later in his text, to speak again about time, this time, an argument with the historian Yerushalmi, about the chance that stands between no future and a possible future: "In naming these doors [of the future], I dream of Walter Benjamin [*je rêve à Walter Benjamin*]. In his *Theses on the Philosophy of History*, Benjamin designates the "narrow door" for the passage of the Messiah, 'at each second'" (1995a, 69). In dreaming of Walter Benjamin, the German Jewish thinker who died during the Holocaust and who has barely had the time to write, at the beginning of the war, his unforgettable *Theses* about the philosophy of history and its relation to futurity, Derrida repeats his return to the German language archive as well as to the Jewish thinkers, Freud and Benjamin, in whose writing he finds the chance of a dreaming of the future, the chance of encountering each second as "the straight gate through which the Messiah might enter" (Benjamin 1969, 264). From following traces to writing in the ashes—this is not only the trajectory of the concept of the archive in its burning-dreaming-conceptualisation. It is also the trajectory of the very figures of these burning conceptualisations, all these authors themselves signs or dreaming figures.

THE FOOTSTEPS OF PSYCHOANALYSIS

"I too am dreaming." (Davoine 2007)

We may indeed perhaps read, by chance, emerging from the dream in Jensen, and repeated in Freud, the mark of a footprint. The first appearance of Gradiva in Pompeii is marked by a single word set off, typographically, in a separate line, a typographical separation that makes an immediate impression: "plötzlich," "suddenly":

Hanold looked before him down the Strada di Mercurio [. . .]
  Then suddenly—
  With open eyes he gazed along the street, yet it seemed to him as if he were doing it in a dream. A little to the right something suddenly stepped forth [. . .] and across the lava stepping-stones, which led from the house to the other side of the Strada di Mercurio, Gradiva stepped buoyantly.

This word, "suddenly," also appears in Hanold's first vision of Gradiva in his dream, and proliferates across Jensen's text and into Freud's after its appearance along with her image. "Suddenly" is, of course, the language of the unexpected, of the accident, of fright, of the event marked by trauma. It is also, here, the word of a chance meeting, an encounter that will also be, in Freud's reading, the beginning of a psychoanalytic process of encounter, or, we could say, the beginning of a certain kind of unprecedented witness. The "suddenly" thus emerges as an accident, mere chance, the beginning, perhaps, of a figure, and a future concept, but one that arises from and must always return to the dream, to the site of chance meetings and to the enigmatic language of the literary. This word, then, emerging from the dream, may be the trace of the trauma, perhaps the footprint of Gradiva, or of psychoanalysis, as it leaves its footprints or as it may simply be disappearing, suddenly, imperceptibly, at the heart of its own burning, in the history of its writing, in the history of "today."[14]

> If you never read this, no one will. [. . .] Or. Or perhaps no. Perhaps these words need the air that is out in the world. Need to fly up then fall, fall like ash over acres of primrose and mallow.[25]

NOTES

1. Jacques Derrida, *Mal d'Archive*, "Prière d'insérer"; this insert is not included in the English translation.
2. On the archaeological metaphor in Freud, see Armstrong (1999); Cassirer Bernfeld (1951); Hake (1993); Kuspit (1989); Kujundz'ic (2004); Larsen (1987); Reinhard (1996); Rudnytsky (1995); Spence (1987); Tögel (1989).
3. Alan Bass (2006) has a good summary of the section of this piece. On *Archive Fever* see also Steedman (2001).
4. Freud thus allows us to think of the potential historicity of the very concept of *Nachträglichkeit*. On the development of this concept see also Laplanche (2006).
5. See Caruth (1996), ch. 3: "Traumatic Departures: Survival and History in Freud."
6. Lifton (1996). See also Caruth (1995).
7. On Freud's reading of Jensen's *Gradiva*, see Bergstein (2003); Downing (2006); Gilman (1993); Hertz (1997); Jacobus (1986); Johnson (2010); Kofman (1991); Lawlor (1998); Møller (1991); Rand and Torok (1997); Rohrwasser et al. (1996). On Derrida and the figure of ash see also Krell (2000).
8. This translation is from the Strachey translation edited by Neil Hertz in *Sigmund Freud: Writings on Art and Literature*, Stanford: Stanford University Press, 1997; translation modified.
9. Perhaps an echo of "verweile doch, du bist so schön," from Goethe's *Faust*.
10. It is interesting to recall Fiorelli's work at the site of Pompeii in the nineteenth century, drilling holes in the tephra, sending plaster down into the cavities created by the incinerated bodies and bringing up their forms. The volcano, in the context of Derrida's reading of Freud's interpretation of Jensen, becomes something of a "writing machine," in the sense in which Derrida refers to it in "Freud and the Scene of Writing." Of course the total incineration of bodies, understood in the context of the total destruction of the archive, would leave no forms or traces behind.
11. It is interesting to note that the word in German is not "foot-print" but "toe-print," which is often lost in translation. As Cynthia Chase pointed out to me, the toe takes on the form of a whole—the foot—as it passes into translation.
12. See also Jacques Derrida, "No Apocalypse, Not Now (full speed ahead, seen missiles, seven missives), tr. Catherine Porter and Philip Lewis, *Diacritics*, Summer 1984.
13. Personal communication.
14. See also Derrida (1984): "The hypothesis of this total destruction watches over deconstruction, it guides its footsteps."

WORKS CITED

Armstrong, Richard H. 1991. "The Archaeology of Freud's Archaeology: Recent Work in the History of Psychoanalysis." *International Review of Psychoanalysis* 13 (1): 16–20.

Bass, Alan. 2006. *Interpretation and Difference: The Strangeness of Care*. Stanford: Stanford University Press.

Benjamin, Walter. 1969. "Theses on the Philosophy of History." In *Illuminations: Essays and Reflections*. Edited by Hannah Arendt. Berlin: Schocken.

Bergstein, Mary. 2003. "Gradiva Medica: Freud's Model Female Analyst as Lizard-Slayer." *American Imago* 60 (3): 285–301.

Caruth, Cathy. 1995. "An Interview with Robert Jay Lifton." In *Trauma: Explorations in Memory*. Edited by Cathy Caruth, 128–50. Baltimore, MD: The Johns Hopkins University Press.

———. 1996. *Unclaimed Experience: Trauma, Narrative and History*. Baltimore, MD: The Johns Hopkins University Press.

Cassirer Bernfeld, Suzanne. 1951. "Freud and Archaeology." *American Imago* 8: 107–128.

Davoine, Francoise. 2007. "The Characters of Madness in the Talking Cure" *Psychoanalytic Dialogues* 17 (5): 627–38.

Derrida, Jacques. 1978. "Freud and the Scene of Writing." In *Writing and Difference*. Translated by Alan Bass, 246–91. Chicago: Chicago University Press.

———. 1984. "No Apocalypse, Not Now (full speed ahead, seen missiles, seven missives)." Translated by Catherine Porter and Philip Lewis, *Diacritics* 14 (2): 20–31.

———. 1991. *Cinders*. Translated, edited and introduced by Ned Lukacher. Nebraska: University of Nebraska Press.

———. 1995a. *Archive Fever: A Freudian Impression*. Translated by Eric Prenowitz. Chicago: Chicago University Press.

———. 1995b. *Mal d'Archive: une impression freudienne*. Paris: Galilée.

———. 2005. "Poetics and Politics of Witnessing." In *Sovereignties in Question: The Poetics of Paul Celan*. Edited by Thomas Dutoit and Outi Pasanen, 65–96. New York: Fordham University Press.

Derrida, Jacques and Elisabeth Weber. 1995. "Passages—From Traumatism to Promise." In *Points . . . Interviews, 1974–1994*. Edited by Elisabeth Weber, 372–98. Stanford: Stanford University Press.

Downing, Eric. 2006. "Archaeology, Psychoanalysis, and Bildung in Freud and Wilhelm Jensen's *Gradiva*." In *After Images: Photography, Archaeology, and Psychoanalysis and the Tradition of Bildung*, 87–166. Detroit: Wayne State University Press.

Freud, Sigmund. 1927. *Delusion and Dream: An Interpretation in the Light of Psychoanalysis of "Gradiva," a novel by Wilhelm Jensen, which is Here Translated*. Translated by Helen M. Downey. New York: New Republic.

Gilman, Sander L. 1993. *Freud, Race, and Gender*. Princeton: Princeton University Press.

Hake, Sabine. 1993. "Saxe Loquuntur: Freud's Archaeology of the Text." *Boundary* 2 (20): 146–73.

Hertz, Neil. 1997. "Foreword." In *Writings on Art and Literature*, Sigmund Freud. Edited by Werner Hamacher and David E. Wellbery, ix–xx. Stanford: Stanford University Press.

Jacobus, Mary. 1986. *Reading Woman: Essays in Feminist Criticism*. New York: Columbia University Press.

Jensen, Wilhelm. 1927. "Gradiva: A Pompeiian Fantasy." In *Delusion and Dream: An Interpretation in the Light of Psychoanalysis of "Gradiva," a novel by*

Wilhelm Jensen, which is Here Translated, Sigmund Freud. Translated by Helen M. Downey, 3–120. New York: New Republic.

Johnson, Barbara. 2010. *Moses and Multiculturalism*. Berkeley: University of California Press.

Kofman, Sarah. 1991. *Freud and Fiction*. Cambridge: Polity Press.

Krell, David. 2000. *The Purest of Bastards: Works of Mourning, Art, and Affirmation in the Thought of Jacques Derrida*. University Park, Pennsylvania: Penn State University Press.

Kujundz'ic, Dragan. 2004. "Archigraphia: On the Future of Testimony and the Archive to Come." *Discourse* 25 (1–2): 166–88.

Kuspit, Donald. 1989. "A Mighty Metaphor: The Analogy of Archaeology and Psychoanalysis." In *Sigmund Freud and His Art*. Edited by Lynn Gamwell and Richard Wells, 133–51. New York: Harry Abrams.

Lacan, Jacques. 1998. "Tuché and Automaton." In *The Four Fundamental Concepts of Psychoanalysis: Book XI of the Seminar of Jacques Lacan*, 53–66. New York and London: Norton.

Laplanche, Jean. 2006. *Problématiques VI, L'Après-coup*. Paris: Presses Universitaires de France.

Larsen, Steen F. 1987. "Remembering and the Archaeology Metaphor." In *Metaphor and Symbol* 2 (3): 187–99.

Lawlor, Leonard. 1998. "Memory Becomes Elektra." *The Review of Politics* 60 (4): 796–98.

Lifton, Robert Jay. 1996. "Survivor Experience and Traumatic Syndrome." In *The Broken Connection: On Death and the Continuity of Life*, 163–78. New York: Basic Books.

Møller, Lis. 1991. *The Freudian Reading: Analytical and Fictional Constructions*. Philadephia: University of Pennsylvania Press.

Morrison, Toni. 2006. *A Mercy*. New York: Vintage.

Rand, Nicholas, and Maria Torok.1997. "A Case Study in Literary Psychoanalysis: Jensen's *Gradiva*." In *Questions for Freud: The Secret History of Psychoanalysis*, 54–75. Cambridge: Harvard University Press.

Rapaport, Herman. 2003. *Later Derrida: Reading the Recent Work*. New York: Routledge.

Reinhard, Kenneth. 1996. "The Freudian Things: Construction and the Archaeological Metaphor." In *Excavations and their Objects: Freud's collection of Antiquity*. Edited by Stephen Barker, 57–80. Albany, NY: State University of New York Press.

Rohrwasser, Michael et al. 1996. *Freuds pomejanische Muse: Beiträge zu Wilhelm Jensens Novelle "Gradiva."* Wien: Sonderzahl.

Rudnytsky, Peter. 1995. "Freud's Pompeian Fantasy." In *Reading Freud's Reading*, Sander Gilman ed, 211–31. New York: New York University Press.

Spence, Donald P. 1987. *The Freudian Metaphor: Toward Paradigm Change in Psychoanalysis*. New York: Norton.

Steedman, Carolyn. 2001. *Dust*. Manchester: Manchester University Press.

Tögel, Christfried. 1989. *Berggasse—Pompeji und Zurück: Sigmund Freuds Reisen in die Vergangenheit*. Tübingen: Diskord.

Yerushalmi, Yosef Hayim. 1991. *Freud's Moses: Judaism Terminable and Interminable*. New Haven: Yale University Press.

2 Apocalypses Now
Collective Trauma, Globalisation and the New Gothic Sublime

Avril Horner

Gothic and trauma theory have flourished since 1980: their ascendancy into our minds and our vocabularies is thus a fairly recent phenomenon that seems to play to our collective sense of feeling beleaguered, endangered and isolated as modern subjects. It is true that only a small minority of people suffer from post-trauma syndrome in the clinical sense of the word—that is, experiencing hallucinations, flashbacks, insomnia and other symptoms; the reaction to trauma that, by definition, as Cathy Caruth has argued, cannot be represented directly and thereby resolved. However, many cultural critics have argued that even ordinary experiences of life within late modernity can become cumulatively traumatic.[1] Such ordinary experiences must now include continual media exposure to violent and terrifying events. As Steven Bruhm has put it, "I may not *be* traumatised at the moment of reading" (or, we might add, of viewing) "but I certainly join with the Gothic mode in *feeling like one who is traumatized*" (2002, 272). Such vulnerability is exacerbated by the viral-like speed with which news about the latest suicide bombing or environmental disaster travels to our screens. In our global world, individuals are both well informed and highly fragile subjects.

In this chapter I shall examine how Gothic apocalypse narratives are used by authors and directors to explore a growing sense of vulnerability in a world now well acquainted with the notion of trauma. As James Berger has noted:

> Apocalypse and trauma are congruent ideas, for both refer to shatterings of existing structures of identity and language [. . .] Post-apocalyptic representations are simultaneously symptoms of historical traumas and attempts to work through them. (1999, 19)

I should make it clear that here I use the word "trauma" not to indicate a clinical condition but in the cultural/historical sense that Berger uses it, that is, to suggest a collective psychological response to a cataclysmic disaster or a profoundly terrifying event. I also use the term "Gothic" in its widest sense—as a mode of excess (Botting 1996, 1) that signifies horror, the abject, the supernatural and the uncanny; the uncanny as formulated by Freud (1985, 339–76) and as developed by recent critics to suggest "a widely used figure for the simultaneous homelessness of the present, and haunting by the past, that has been associated with modernity since at least the time of Baudelaire" (Collins and Jervis 2008, 2).

It is no coincidence that scholars of the Gothic enthusiastically embraced Freudian theory during the 1980s; in a sense both the Gothic and psychoanalysis tell the same story about the human condition. Indeed, they share the same discursive tropes: secrets, dark spaces, hauntings and encryption. Evident across art, literature, music and popular culture, Gothic can be seen as a "hybrid assemblage" that evolved in response to a growing anxiety about "the subject in a state of deracination, of the self finding itself dispossessed in its own house, in a condition of rupture, disjunction, fragmentation" (Miles 2002, 3). The narratives of psychoanalysis, trauma and the Gothic all emphasise the human tendency to the irrational and to the death-drive. However, Gothic fictions go further in so far as they pit horror and the uncanny against the Western optimistic Enlightenment agenda of progress through science, reason and objectivity. Indeed, the Gothic challenges both conventional history and the integrity of the modern subject who is seen as caught in what Alexandra Warwick has wittily described as "the *unheimlich* manoeuvre [. . .] in which the position of the subject collapses from the illusion of coherent dominance into fragmentary dissolution" (1999, 82). Gaps and aporias characterise Gothic texts and trauma narratives, which are both also marked by repetition and return, fragmentation and split subjectivities. Trauma and Gothic narratives equate to disruption, irruption and melancholy. It is therefore not surprising to find that they frequently seep into each other: the Gothic mode feeds off the horror of traumatic situations for plot while trauma narratives are often structured through Gothic effects in order to convey a sense of haunting, of matters unresolved, of the past intruding into the present—and of narrative itself fragmenting or becoming unreliable.

Gothic and trauma theory became ubiquitous in the run up to the millennium. In 1974, Angela Carter remarked that "We live in gothic times" (1974, 122) and in 1980 David Punter published *The Literature of Terror* (1996) which established Gothic as a respectable academic field of study. In 2007 Alexandra Warwick noted that the Gothic is everywhere:

> From academic publishers' catalogues to *Vogue*, university classrooms to internet chat rooms, television and film listings to art exhibitions and music festival line-ups [. . .] Gothic can no longer proceed from the margins, because there is no marginality, it is where everybody wants to live. Normality is Gothic and Gothic is normal, both in criticism and contemporary culture. (2007, 5, 14)

In a slightly different vein, Roger Luckhurst, in his recent book *The Trauma Question* (2008a), explores the nature and extent of "trauma culture" from 1980 and notes its subsequent pervasiveness. Like the Gothic, the concept of trauma has been culturally processed so as to emerge in manifold forms. On the one hand, in the wake of Cathy Caruth's groundbreaking work, intelligent academic analyses of how history, trauma and memory relate to each other are represented by books such as Michael Rothberg's *Multidirectional Memory: Remembering the Holocaust in the Age of Decolonization* (2009). On the other hand, in popular culture the word "trauma" is now bandied about with impunity almost as if it endows the modern subject with a mark of authenticity. Celebrity biographies, for example, frequently include descriptions of a "traumatised" childhood or adolescence. We are thus faced with a spectrum of meaning in which the word "trauma" signifies, at one end, an experience beyond articulation and outside the boundaries of normal experience and, at the other end, an aspect of "normality" itself. In short, trauma—like Gothic—is everywhere. Applied to whole nations, it has become a defining focus in film studies, as in Adam Lowenstein's *Shocking Representations: Historical Trauma, National Cinema and the Modern Horror Film* (2005) and Linnie Blake's *The Wounds of Nations: Horror Cinema, Historical Trauma and National Identity* (2008).

The contemporaneous rise of Gothic and of trauma theory must owe something to the fact that in our globalised world we live vicariously and constantly with the aftermath of horrifying events. Since New York 9/11/2001, Madrid 3/11/2004, London 7/7/2005 and subsequent terrorist attacks across the world, citizens are haunted on an almost daily basis by particular scenarios, whether in the form of follow-up official enquiries, individual testimonies, or fictional representations. In this sense, "collective trauma" is now endemic to the modern world and is exacerbated by the fact that we live in globalised societies based on risk culture. For Ulrich Beck, as Luckhurst points out, the risk society "is a catastrophic society. In it, the exceptional condition threatens to become the norm" (2008a, 213–14). In response to this brave new world, we are now seeing the emergence of Global Gothic—that is, Gothic films and fictions arising from fears about the impact and effects of globalisation (including terrorist acts) on cultures, societies and individuals. I suggest that while in some ways the terms "Gothic" and "trauma" have become partially normalised within Western culture, the recent upsurge in apocalyptic and post-apocalyptic fictions reflects a state of collective trauma endemic to globalised societies. Of course, given the logical impossibility of anyone being left to document the event after a total apocalypse, such works enact, as Derrida has commented (when writing of projected nuclear warfare and the rhetoric that embodies it) "a fabulous specularization" and must remain "fabulously" (in the original sense of the word) textual (Derrida 1984, 23).[2] However, while purporting to be fantastic projections of the future, such fictions have a valuable cultural function in that they enable a continued engagement with past events and current fears, albeit at an aesthetic remove.

The apocalypse narrative is not new of course: the Book of Revelations left a strong mark on Western literature, one of the most famous examples being Mary Shelley's *The Last Man*, published in 1826. Apocalyptic fictions and predictions, like the Gothic, tend to bubble up during times of crisis; it is, for example, generally accepted that both the Second World War and the Cold War resulted in spates of apocalyptic writing. Over the last twenty or so years, however, apocalyptic scenarios have become increasingly common in both film and fiction, particularly evident in the remaking of former apocalyptic tales, for example the three film versions of Richard Matheson's 1954 novel *I am Legend* released in 1964, 1971 and 2007 and the sequel to Danny Boyle's 2002 film *28 Days Later*, released in 2007 as *28 Weeks Later*. The Gothic often plays an important role in the creation of such worlds, particularly in the revival of the Gothic monster. In Danny Boyle's *28 Days Later*, for example, the figure of the zombie is horribly revived. In this British film, people infected with the "rage" virus—inadvertently released from a Cambridge laboratory by a group of animal activists—become dehumanised and pathologically aggressive. Their murderous rage, it has been suggested, represents the dehumanising effects of a post-Thatcher, post-industrialised fast-moving society. The "infected," as they are known, run, leap and attack from nowhere: in recent films, as Fred Botting has observed, "Zombies, slow, lumbering, relentless and modern, are replaced by 'zoombies', fast-moving figures of a fast-food culture and fast camera cinema," suggesting for some the almost manic hyperproductive bodies needed by corporate enterprises, for others the internal rage generated by such a life-style (Botting in Phillips and Witchard 2010, 158, 159). Such a critique of global corporatism is also evident in the film *28 Weeks Later*, released in 2007, which foregrounds as its setting London's Isle of Dogs and Canary Wharf, an area described by one critic as the "most public and visual expression of 1980s aggressive monetarism" (Bird 1993, 123). This sequel to the earlier dystopia is even darker: whereas the ending of *28 Days Later* has the Irish hero, his female companion and a young girl (the archetypal family) fleeing to the rural safety of the Lake District where they will be rescued by the Finnish Air Force, *28 Weeks Later* ends with a helicopter leaving for France, a supposedly uninfected country. The film's closing shots, however, show a crashed helicopter and infected hordes teeming round the Eiffel Tower. While envisioning the end of the world as we know it, these apocalyptic narratives engage—in a displaced manner—with images that have haunted the recent past. For example, the disjointed shots of London being firebombed at night in *28 Weeks Later* eerily recall the jerky newsreel footage of Baghdad being invaded in 2003 at the start of the war in Iraq. The future as imagined in *28 Weeks Later* can thus be read as an alternative representation of Western military intervention in the Middle East, albeit one that is neither sanitised nor censored and that shows the true human cost of the strategy in images of flailing burning bodies. For Kirk Boyle, scenes in the film of detained Arab men being tortured at the internment camp at Bexhill evoke images "reminiscent of Abu Ghraib and Guantanamo" (2009). To use Adam Lowenstein's phrase, such scenes provide an "allegorical moment" in films—a moment which exists "as a mode of confrontation, where representation's location between past and present, as well as between film, spectator, and history demands to be recalibrated" (2005, 12).

Most recent apocalypse fictions fall into one or more of three categories: "plague" narratives; environmental disasters; and war narratives relating to either terrorism or nuclear explosions. Such terrors are often implicitly associated with globalisation, a recent economic phenomenon that—like the Gothic itself—erodes the boundaries that shore up our identity Indeed, Žižek asserts unequivocally that:

> the global capitalist system is approaching an apocalyptic zero-point. Its "four riders of the apocalypse" are comprised by the ecological crises, the consequences of the biogenetic revolution, imbalances within the system itself (problems with intellectual property; forthcoming struggles over raw materials, food and water), and the explosive growth of social divisions and exclusions. (2010, x)

Given this state of impending global crisis, it is not surprising that post-millennial apocalypse narratives invariably reflect or anticipate actual events: the terrorist attacks of the last twelve years; the tsunamis that hit Thailand in 2007 and Japan in 2011, the Florida oil leak in 2010, and media panics surrounding bird-flu, swine-flu and the SARS virus. Flashed onto our screens, such events can make the world itself seem hostile and unheimlich; panic becomes a default reaction. It is therefore not surprising to find Gothic devices used in recent apocalyptic narratives: paranoia is, after all, the grace note of Gothic. In enabling a mediated engagement with past horrors as well as imagined fears, Gothic apocalypses are both retrospective and futuristic. Extrapolating from past events that traumatised many, they seek to represent not just what has happened to some but what could happen to all.

P. D. James' novel, *The Children of Men*, published in 1992, provides a good point of comparison with post-millennial apocalypse narratives. Set in England in 2021, its plot is charted through the diary of an Oxford don, Dr Theodore Faron. We learn that since 1995, which marked the last human birth, the world has been beset by mass infertility, the cause of which is unknown. The last survivor of the Omegas—the name given to those born during 1995—has just died and the human race seems doomed to extinction. Like everyone else, Theo suffers from what is described in the novel as "universal anomie" (James 1994, 143) but his life changes when he is approached by a woman called Julian, who has miraculously become pregnant and who is a member of a small dissident group called the Five Fishes. Finding themselves at risk and likely to be tracked, the group decides to search for a place of quiet safety where the child can be born. Eventually they manage to reach a wood-shed deep in Wychwood Forest in Oxfordshire. Soon after the baby—a boy—is born, the Warden, who autocratically rules the country and who also happens to be Theo's cousin, arrives. Theo shoots him through the heart, removes the ring of government from his finger, and places it on his own. The novel ends with a cameo hauntingly reminiscent of the Holy Family—Theo, Julian and the baby in the woodshed—and with the council and the Warden's private army waiting outside for their new orders.

The Children of Men clearly represents some of the dominant anxieties of the early 1990s: the burden of an increasingly ageing population (a ritual known as the "quietus" forces the old and infirm to commit suicide in group drownings); a concern that the use of modern chemicals in everyday life might be causing fertility problems; anxiety about the treatment of immigrants; the evolution, on the one hand, of a Thatcherite wealthy and powerful elite and, on the other, of a disenfranchised population easily manipulated by that elite. Like all good writers of dystopian fictions, James sensed these fears and exaggerates them. Significantly though, the film based on the book, entitled *Children of Men*, which was directed by Alfonso Cuarón and released in 2006, is set very differently. The pastoral and redemptive world of James' novel is here replaced by a devastated London and derelict buildings in Bexhill-on-Sea in Sussex, where an uprising in 2027 against the authorities has resulted in chaos, murder and mayhem. The dark cityscapes of the film are typical of the urban Gothic tradition in which the recurring trope of the decaying city seems to suggest real fears about the nature and survival of "civilisation." The holy trinity of father, mother and child is preserved in the film, however. The last scene shows the trio drifting offshore in a tiny boat in a Turner-like seascape, Theo (the Joseph figure) dying from gunshot wounds while Kee (a black refugee) nurses her newly-born daughter.[3] The emergence from the mist of a ship called "Tomorrow" owned by the Human Project, a dissident community based on the Azores, suggests that the human race will survive after all, for—as Berger notes—"Very few apocalyptic representations end with the End" (1999, 34). Somewhat sentimental perhaps, this closure nevertheless interestingly recalls the rise of the Greenpeace movement during the 1990s; the film's message is more obviously ecological than that of the novel, reflecting perhaps a growing anxiety about economic systems, climate change and the environment. Indeed, Kirk Boyle argues that "by yoking together images of seemingly disconnected crises over the course of 109 minutes (images of globalisation, immigration, inequity, environmental degradation, permanent statues of emergency, politics of fear, surveillance society, terrorism, and ghettoes), Cuarón argues for their dialectical relationship" (2009).

I want now to turn briefly to three post-millennial apocalyptic novels that use Gothic effects to address, respectively, the aftermath of a terrorist attack, a nuclear winter and an environmental disaster. Chris Cleave's novel, *Incendiary*, structured through a first person narrative that takes the form of an outpouring to Osama Bin Laden, was published in 2005. When the novel opens, the narrator, a nameless working-class woman from the East End of London is married to a policeman, a bomb disposal expert, and they have a four-year-old son. Husband and son—or "the chaps" as the woman calls them—are both killed, along with over 1,000 other spectators, in an al Qaeda terrorist attack on a football ground in London where they have gone to watch Arsenal play Chelsea. Reaching the ground, the narrator is faced with a scene of Gothic horror in which piles of bodies are surrounded by body parts and the earth is sticky with blood. The rest of the novel vividly portrays the narrator's traumatised state; the horrific deaths of her husband and child—identifiable only by their teeth—render her subject to nightmares, hallucinations and the shakes. Unable to work through her son's death, she constantly sees him alive, by her side. He is both her son and a strangely uncanny being with flame flickering round his ginger hair crying out to her for help. Indeed, it is arguably the vivid fictional portrayal of post-traumatic stress disorder that makes this novel so memorable. In the words of the narrator:

> I always thought an explosion was such a quick thing but now I know better. The flash is over very fast but the fire catches hold inside you and the noise never stops. You can press your hands on your ears but you can never block it out. The fire keeps on roaring with incredible noise and fury. (Cleave 2005, 168)

These words communicate something of the paradoxical nature of trauma: as Cathy Caruth has noted: "The flashback or traumatic re-enactment conveys [. . .] both the truth of the event, and the truth of its incomprehensibility" (1995, 153). *Incendiary* closes with scenes of urban breakdown in which the narrator—who now works as a shelf-stacker—pleads with Osama Bin Laden to "stop blowing the world apart" (242). The novel is book-ended by two quotations from the inscription on the Monument to the Great Fire of London which vividly commemorate the destruction of the city in 1666: "a most terrible fire broke out, which not only wasted the adjacent parts, but also places very remote, with incredible noise and fury." The novel's last five words echo this inscription: "Come to me," the narrator writes to Bin Laden, "and we will blow the world back together WITH INCREDIBLE NOISE AND FURY." This plea, criticised by some reviewers as sentimental and unconvincing, in fact registers both the incomprehensibility of the event for the narrator and the way it has seared itself into her language; like many trauma survivors, she has become possessed by the event that traumatised her. The novel's closure also signals a sense of history as cyclical, as a series of quasi-apocalyptic events rather than a linear progression towards End and Revelation. Recognised as a brilliant first novel, *Incendiary* was nevertheless eclipsed by the grisly coincidence of its publication on the very day—7 July 2005—that real terrorists bombed three London underground trains and a bus, killing more than fifty and injuring hundreds. Its mixture of demotic black humour, social satire and gothic horror were simply too close for comfort in the face of an actual terrorist attack. In particular, many found its vision of London teetering on the edge of apocalypse, with a midnight curfew, barrage balloons around the city, and another bomb threat resulting in panic and riot, just too uncomfortable to contemplate in July 2005, a month of real panic, terror and grief for Londoners. However, Cleave's use of Gothic effects to suggest the horrors of such a situation resonates beyond British society. Floating down the Thames on an upturned boat in order to avoid the inferno which London has become after the riots, the narrator closes her eyes to avoid seeing "all the bodies floating down the river with me." But then:

> Once when I did open my eyes it was hours later and I was going under Southwark Bridge. [. . .] It was a seagull squawking that made me open my eyes. There was an Asian boy maybe 16 or 17 years old floating between my boat and the sunset. The boy was 2 feet from me he was floating face up in a McDonald's uniform. Grey polyester trousers maroon short-sleeved shirt and a maroon baseball cap. The seagull was sticking his head in under the peak of the baseball cap to eat the boy's left eye. The boy had a name badge it said HI my name is NICK how can I help you today? He had 2 out of 5 merit stars on his badge and they glistened in the sunset. (224)

Cleave here implicitly links individual tragedies such as the deaths of a British Asian boy and the nameless narrator's son and husband with the workings of a global capitalist economy which renders its workers mere cogs in the machinery of corporatism and politics: "All the violence in the world is connected it's just like the sea" (10) thinks the woman at one point. It is no coincidence of course that the narrator lives in working class Bethnal Green, an area not far from London's Canary Wharf—the city's hub of global financial deals. As in *28 Weeks Later*, the new global order has turned the city into a microcosm of first world and third world inhabitants, Canary Wharf representing the power of deregulated free financial currents while the working class and the unemployed live restricted lives in nearby impoverished areas. In this respect, the novel anticipates the argument of Naomi Klein's *The Shock Doctrine: The Rise of Disaster Capitalism*: that the social effects of corporate globalisation will include planned segregation, with an increasing number of protected living areas for the rich, known as "Green Zones" (2007, 420–22), while the living conditions of the poor become worse.

In Cormac McCarthy's *The Road* (2006), there are no inhabitants, only wandering survivors. In this extraordinary reworking of the American road novel McCarthy portrays a ruined earth in which a father and son, referred to only as "the man" and "the boy," struggle to live. In what appears to be a nuclear winter, America has been reduced to a landscape of fire and ashes, lashed by rain and snowstorms, with roaming bands of marauders resorting to murder and cannibalism in order to survive. In another—rather more subtle—appropriation of the Gothic zombie figure, most survivors have become the undead, "Slouching along with their clubs in their hands" (McCarthy 2007, 62–63). Indeed, the man remembers his wife—who, we learn, committed suicide rather than live in such a world—having described them as "the walking dead in a horror film" (57). Although the journey to reach the coast—the quest narrative—pushes the story forward, the past continually disrupts the present in the form of the father's dreams, nightmares and memories. Some of these evoke a natural world and a faith now destroyed and associated therefore with the spectral and the lost: a snowflake expires in his hand "like the last host of Christendom" (15) and in a dream the "uncanny taste of a fresh peach" evokes "some phantom orchard fading in his mouth" (17). Social structures have been replaced by "deranged chanting [. . .] the screams of the murdered (and) [. . .] the dead impaled on spikes along the road" (33). In one grand and apparently abandoned house, the man discovers, in a storage cellar beneath the pantry, a group of people kept alive so that their bodies can serve as food for a small and chillingly well-organised band of survivors:

> Huddled against the back wall were naked people, male and female, all trying to hide, shielding their faces with their hands. On the mattress lay a man with his legs gone to the hip and the stumps of them blackened and burnt. The smell was hideous.
>
> Jesus, he whispered.
>
> Then one by one they turned and blinked in the pitiful light. Help us, they whispered. Please help us. (116)

It is no coincidence that the grand house in which the band of cannibals keeps their human prey was built and maintained through the exploitation of others: "Chattel slaves had once trod those boards bearing food and drink on silver trays" (112). This house is haunted, like America itself, by its past deeds of exploitation. At one stop on their journey, exploring a looted drugstore, the man and boy come across a "human head beneath a cakebell at the end of the counter" (195); at another, they see a small group of adults roasting a dead baby on a spit. In *The Road*, an austere and lyrical work, such Gothic horrors are not gratuitous: they are presented as a test of the two main characters' ability to maintain some sort of ethical awareness in the hell that their homeland has become. The father's wisdom and stoicism is offset by his son's uncanny awareness and natural altruism; they define themselves as two of "the good guys" and, indeed, their refusal to participate in cannibalism signifies one ethical dimension of what it means to be human. They are the walking living rather than the walking dead, seeing themselves as "carrying the fire" (87). Their faith in human nature is confirmed by the family group who adopt the boy after his father's death. This rescue-closure, typical of apocalyptic narratives and, again, hauntingly reminiscent of the Holy family, is, however, counter-pointed by a final paragraph that, both mystical and elegiac, presents nature as a sublimely ancient and transcendent force:

> Once there were brook trout in the streams in the mountains. You could see them standing in the amber current where the white edges of their fins wimpled softly in the flow. They smelled of moss in your hand. Polished and muscular and torsional. On their backs were vermiculate patterns that were maps of the world in its becoming. Maps and mazes. Of a thing which could not be put back. Not be made right again. In the deep glens where they lived all things were older than man and they hummed of mystery. (307)

This cameo of the natural world, unrepresented in the film version of the novel, recalls Blake's mystical vision in "To see a World in a Grain of Sand" ("Auguries of Innocence"). McCarthy (a lapsed Catholic) seems to be prompting his reader to think about apocalypse in relation to its original meaning—a revelation, a lifting of the veil, a disclosure of truth—in an era dominated by notions of economic progress and global supremacy. "Apocalyptic desire," as Berger notes, is both "a longing for the end" and "a longing also for the aftermath, for the New Jerusalem," for change and transformation (Berger 1999, 34). McCarthy's evocation of the numinous and revelatory here indicates a post-Romantic focus on the natural world that combines modern ecological awareness with a quasi-religious sensibility. Such a vision can be easily caricatured as "a fundamentalist religion adopted by urban atheists looking to fill a yawning spiritual gap plaguing the West" (Morton 2009, 4 quoted in Žižek 2010, 341) but the huge critical and popular success of *The Road* suggests that, like the figure of Death on a Pale Horse in the eighteenth century, it articulates common fears and nightmares. In her recent book *Gothic Riffs: Secularizing the Uncanny in the European Imaginary 1780–1820*, Diane Long Hoeveler's examines the reiterative work of the Gothic in writing and other forms during this period, suggesting that:

> rather than force people to choose exclusive allegiance to either the immanent order or the transcendent, the rise of ambivalent secularization actually allowed modern Europeans to inhabit an imaginative space in which both the material (science and reason) and the supernatural (God and the devil) coexisted as equally powerful explanatory paradigms. This uneasy coexistence of the immanent and the transcendent can be seen throughout the gothic corpus. (Hoeveler 2010, 6)

This "uneasy coexistence of the immanent and the transcendent" perhaps accounts for the fragile survival of the belief or hope we find in some texts that the apocalypse might just usher in a New Jerusalem, a new order. However, this utopian flipside to the dystopian underside is more often missing from, or at least muted within, many recent apocalyptic fictions. As Bull has noted "We seem to be in the presence here of a debased millenarianism without a compensating utopian vision" (1995, 212).

Liz Jensen's *The Rapture* (2009), which has at its heart fears about an environmental disaster and the rise of evangelicalism and fundamentalism in a global world, offers no "compensating utopian vision" although it does finally evoke the numinous. A resolutely bleak novel, it offers just a glimmer of hope at the end in the pregnancy of the narrator, a psychotherapist (herself traumatised by a car accident that has left her paralysed from the waist down), who, with her partner, survives the end-of-the-world scenario through being airlifted by helicopter from London. Described by one critic as an "original blend of contemporary Gothic, zeitgeist anxiety and scientific possibility" (inside cover), its apocalyptic closure takes the form of a global tsunami triggered by methane fires resulting from the collapse and implosion of an oil rig off the coast of Norway—a fictional disaster which eerily anticipated the Gulf of Mexico oil slick catastrophe in 2010 caused by a British Petroleum rig operation accident. When the novel opens, the narrator is treating a disturbed adolescent girl, Bethany, who was abused by her father, an evangelical Christian, and who murdered her mother at the age of fourteen by ramming a screwdriver through her eye-socket into her brain. Like many abused characters in fiction, Bethany has an uncanny ability to see into the future.[4] The novel closes with the narrator and a few others watching the destruction of the world from the helicopter which has rescued them. Bethany is next to her, and is strangely exultant about having foreseen and predicted the end of the world:

> The fire spreads greedily as though devouring pure oil, yellow flames bursting from the crest of the liquid swell, triggering star-burst gas explosions above. With a deep-throated bellow the wave gushes across the landscape, turning buildings and trees to matchwood in an upward rush of spume. As the force catapults us upward, the scene shrinks to brutal eloquence: a vast carpet of glass unrolling, incandescent, with powdery plumes of rubble shooting from its edges, part solid, part liquid, and part gas—a monstrous concoction of elements from the pit of the Earth's stomach. There's a gentle pliant crunching and far below buildings buckle, ploughed under, then vanish in the suck. Only a few skyscrapers stand proud of the burning waterscape as the land is relentlessly and efficiently erased. The heat is unbearable, as though the sun itself had plunged into the water and is irradiating us from below. It's almost impossible to breathe. There's a stench of burnt wood, melted plastic, of meat and seafood boiled to the bone. Tiny rainbows dance

across the open side of the helicopter above the pulsing floodwater. It is the most terrible thing I have ever seen.

"It's wonderful," says Bethany. She is staring at it, mesmerised.

"You'll remember it for ever. You'll remember me too. I know you will."

The strange light makes her face look as translucent and ghostly as rice paper. (Jensen 2009, 339–40)

The novel ends with Bethany throwing herself from the helicopter, cartwheeling "down through the vapour," becoming "a comma and then a speck. And then a burning shard, gulped into the abyss" (340). The narrator, however, survives, knowing she will raise her child in a "world I want no part of. A world not ours." There is a clear echo here of the nineteenth-century apocalyptic sublime which frequently manifested itself in the diluvian. As Alexandra Warwick has pointed out, in many Victorian paintings of the apocalypse, the end of the world is caused by a watery deluge: "There are obvious Biblical origins for this image, but it seems to be a hybrid of the sublime landscape and the apocalyptic sublime, with divine forces breaking through nature" (Warwick in Byron and Punter, 1999, 79).

The endings of McCarthy's *The Road* and Jensen's *The Rapture* work to similar effect, reviving a sense of the Gothic sublime in their combination of the terrible and the beautiful. Although McCarthy's novel ambiguously engages with biblical language and with the concept of God, Jensen's novel is resolutely secular. Nevertheless, they share a common feature: at the end of each tale, the horror of nihilism and annihilation is held in tension with a sense of the transcendent or the numinous. We might call this the New Gothic Sublime. This effect is altogether different from the gentle, quasi-religious epiphanies of P. D. James' novel, in which Theo and his two companions see in Wychwood Forest "a white shape followed by another [. . .] It was a deer and her fawn" (310). Their medieval dream-vision experience gives way to a re-visioning of the world in which nature is seen afresh and as benign, creating a very different effect from the last words of both *The Road* and *The Rapture* which portray nature as a primeval and transcendent force. It is indeed as if the Gothic has returned to one of its original aspects, the sublime. Conceptualised by Edmund Burke in 1757 and evident in the work of many Romantic writers and painters, the sublime was understood as arousing a strange mixture of awe, dread and pleasure in the spectator. Indeed, for Burke, the sublime was linked to a mixture of pain and delight, to terror and danger:

Whatever is fitted in any sort to excite the ideas of pain, and danger, that is to say whatever is in any sort terrible, or is conversant about terrible objects, or operates in a manner analogous to terror, is a source of the sublime; that is, it is productive of the strongest emotion which the mind is capable [. . .] When danger or pain press too nearly, they are incapable of giving any delight, and are simply terrible; but at certain distances, and with certain modifications, they may be, and they are delightful [. . .] pain and terror [. . .] are capable of producing delight; not pleasure, but a sort of delightful terror [. . .] Its object is the sublime. Its highest degree I call astonishment. (Clery and Roberts 2000, 113, 121)

We might think here also of more recent formulations of the sublime: for example, Lyotard who, in responding to Kant's work, sees the experience of the sublime as revealing the limits of our conceptual powers, an inadequacy which prompts a sense of the numinous as well as evoking terror. Now, at least in the secular West, the numinous seems to be appearing in the most unlikely places, including popular apocalyptic texts. It might just suggest a growing resistance to a current powerful global economic assumption—that everything, including nature itself, can be reduced to an economic value (one could cite here environment banking and carbon offsets). The New Gothic Sublime seeks to challenge, in short, the dynamics and progress of globalisation.

The texts I have discussed in this chapter are hybrid works. They are all apocalyptic but they can also justly be described as Gothic in the broadest sense of the word—even though they contain no ghosts or spirits—since they engage heavily with horror, the abject and the uncanny. They aim to shake the reader out of a complacent belief that perpetual economic growth and globalisation must inevitably be a good thing; in that sense they revive the anti-Enlightenment agenda of the traditional Gothic novel, raising disturbing questions about the nature of progress and envisioning, with horrible clarity, what might happen to a risk society that takes one risk too many. They feature scenes of urban devastation and ruined nature as Gothic horror scenarios and they imaginatively map various experiences of trauma. These works are part of what we might now call Global Gothic: that is, a creative response to terrorism, continual warfare and environmental disasters resulting from global tensions and global projects, disasters that know no state or geographical boundary—a response to a world in which human relationships and nature itself seem threatened by the demands and ravages of corporate capitalism and new technologies. Taken together, they offer—to use Kirk Boyle's words—"a coherent narrative of globalisation and its discontents" (2009). Whereas individual trauma narratives are difficult and painful retrievals of an encrypted past, apocalyptic fiction is a wild imagining of the future inspired by present or past terrors. For Žižek the recent resurgence of apocalyptic fictions indicates a "temporal reversal—wherein the symbolic depiction precedes the fact it depicts, history as story precedes history as real event—(and) is an indicator of the condition of late modernity in which the real of history assumes the character of a trauma" (2010, 316). Setting the narrative in years to come, the author or director who works with apocalyptic tales is able to offer an alternative reading of history and to bypass the heavy weight of moral responsibility that burdens those seeking to represent actual traumas. The subtlety, circularity, remorse and moral scrupulousness of works such as Anne Michael's *Fugitive Pieces* (1997) and W. G. Sebald's *Austerlitz* (2001) are missing from most popular apocalyptic fictions. However, this does not mean that such fictions should be dismissed as crude or exploitative representations of collective trauma. Indeed, it may well be that the highly sensational content of these works is a vital ingredient in the audience's ability to engage with, and imaginatively negotiate, trauma. As we have seen, such popular genres frequently embrace Gothic effects; perhaps, as Elisabeth Bronfen has claimed, "The Gothic mode of warning is implicitly the only way the truth can be told in a situation of catastrophe" (2009). However, the apocalyptic fictions discussed here seek to move beyond trauma and Gothic horror: engaging imaginatively with the prospect of the world's end, they offer subtexts that encourage an ethics of global responsibility via a retrieval of the numinous and the sublime.

NOTES

1. Cf. Walter Benjamin's comment that Baudelaire "indicated the price for which the sensation of the modern age may be had: the disintegration of the aura in the experience of shock" (Benjamin 1970, 195). See also Mark Seltzer's claim that "the modern subject has become inseparable from the categories of shock and trauma" (quoted in Luckhurst 2008b, 132).
2. See Christopher Norris (1995, 241–47) for an interesting defence of Derrida's essay as an attempt to deconstruct the rhetoric of nuclear warfare, "to take full account of its 'performative' aspect, before it achieves the referential status of a discourse whose final guarantee would be catastrophe itself" (247).
3. See Kirk Boyle (2009) for a more detailed exploration of religious references in the film.
4. See Roger Luckhurst (2008b, 130): "Freud's close colleague Sandor Ferenczi believed in 'the sudden, surprising rise of new faculties after a trauma'. Trauma 'makes the person in question [. . .] more or less clairvoyant', Ferenczi argued, because the passage through trauma was a little death, and thus related to 'the supposition that the instant of dying [. . .] is associated with that timeless and spaceless omniscience'." Other fictional examples of such "new faculties" following the experience of trauma include the heroine's ability to levitate in Barbara Comyns's *The Vet's Daughter* (1959) and the son's ability to see into the future in Stephen King's *The Shining* (1977).

FILMS:

Twenty Eight Days Later. 2002. Directed by Danny Boyle.
Children of Men. 2006. Directed by Alfonso Cuarón.
Twenty Eight Weeks Later. 2007. Directed by Juan Carlos Fresnadillo and Danny Boyle.

WORKS CITED

Benjamin, Walter. 1970. "On Some Motifs in Baudelaire." In *Illuminations*, edited by Hannah Arendt; translated by Harry Zohn, 157–96. London: Jonathan Cape.

Berger, James. 1999. *After the End: Representations of Post-Apocalypse*. Minneapolis: University of Minnesota Press.

Bird, Jon. 1993. "Dystopia on the Thames." In *Mapping the Futures*, edited by Jon Bird, 121–37. London and New York: Routledge.

Blake, Linnie. 2008. *The Wounds of Nations: Horror Cinema, historical trauma and national identity*. Manchester: Manchester University Press.

Botting, Fred. 1996. *Gothic*. London and New York: Routledge.

Botting, Fred. 2010. "Zoombie London: Unexceptionalities of the New World Order." In *London Gothic*, edited by Lawrence Phillips and Anne Witchard, 153–71.

Boyle, Kirk. 2009. "'*Children of Men* and *I Am Legend*': the disaster-capitalism complex hits Hollywood." In *Jump Cut: A Review of Contemporary Media* 51. Accessed August 10 2011. http://www.ejumpcut.org/archive/jc51.2009/ChildrenMenLegend/index.html.

Bronfen, Elisabeth (2009). "Gothic Wars—Media Lust." Lecture given at the 9th Biennial International Gothic Association conference at Lancaster University on 26th July 2009. Accessed March 3 2011. http://www.bronfen.info/index.php/writing/38-writing/135-Gothic.

Bruhm, Steven. 2002. "Contemporary Gothic: Why We Need it." In *The Cambridge Companion to Gothic Fiction*, edited by Jerrold E. Hogle, 259–76. Cambridge: Cambridge University Press.

Bull, Malcolm, ed. 1995. *Apocalypse Theory and the Ends of the World*. Oxford: Blackwell.

Burke, Edmund. 1757. *A Philosophical Enquiry into the Origin of Our Ideas of the Sublime and Beautiful* Vol.I .vii and Vol.IV.vii as cited in Clery and Miles, 2000.

Byron, Glennis, and David Punter, eds. 1999. *Spectral Readings: Towards a Gothic Geography*. Basingstoke: Macmillan Press.

Carter, Angela. 1974. "Afterword" to *Fireworks: Nine Profane Pieces*. London: Quartet Books.

Caruth, Cathy, ed. 1995. *Trauma: Explorations in Memory*. Baltimore and London: The Johns Hopkins University Press.

Cleave, Chris. 2005. *Incendiary*. London: Chatto and Windus.

Clery E. J., and Robert Miles, eds. 2000. *Gothic Documents: A Sourcebook 1700–1820*. Manchester: Manchester University Press.

Collins, Jo, and John Jervis, eds. 2008. *Uncanny Modernity: Cultural Theories, Modern Anxieties*. Basingstoke: Palgrave Macmillan.

Comyns, Barbara. (1959) 1981. *The Vet's Daughter*. London: Virago.

Derrida, Jacques. 1984. "No Apocalypse, Not Now (Full Speed Ahead, Seven Missiles, Seven Missives)." *Diacritics* 14 (2, Special Issue "Nuclear Criticism"): 20–31.

Freud, Sigmund. 1985. "The 'Uncanny'." In *Art and Literature: Jensen's 'Gradiva,' Leonard Da Vinci and other works*. Penguin Freud Library Vol.14, translated by James Strachey, edited by Albert Dickson. Harmondsworth: Penguin.

Hoeveler, Diane Long. 2010. *Gothic Riffs: Secularizing the Uncanny in the European Imaginary 1780–1820*. Columbus: Ohio State University Press.

James, P.D. 1992; 1994. *The Children of Men*. London: Penguin Books.

Jensen, Liz. 2009. *The Rapture*. London: Bloomsbury.

Klein, Naomi. 2007. *The Shock Doctrine: The Rise of Disaster Capitalism*. London: Penguin Books.

King, Stephen. (1977) 1978. *The Shining*. London: Hodder & Stoughton.

Lowenstein, Adam. 2005. *Shocking Representations: Historical Trauma, National Cinema and the Modern Horror Film*. New York: Columbia University Press.

Luckhurst, Roger. 2008a. *The Trauma Question*. London and New York: Routledge.

———. 2008b. "The Uncanny After Freud: The Contemporary Trauma Subject and the Fiction of Stephen King." In Collins and Jervis 2008, 128–45.

Lyotard, Jean-Francois. 1994. *Lessons on the Analytic of the Sublime (Kant's Critique of Judgment, Sections 23–29)*. Translated by Elizabeth Rottenberg. Stanford: Stanford University Press.

McCarthy, Cormac. 2006. *The Road*. London: Picador.

Michaels, Anne. (1997) 1998. *Fugitive Pieces*. London: Bloomsbury.

Miles, Robert. (1993) 2002. *Gothic Writing 1750–1820*. Manchester: Manchester University Press.

Norris, Christopher. 1995. "Versions of Apocalypse: Kant, Derrida, Foucault." In *Apocalypse Theory and the Ends of the World* edited by Malcolm Bull, 227–249. Oxford: Blackwell.

Phillips, Lawrence and Anne Witchard, eds. 2010. *London Gothic*. London: Continuum.

Punter, David. (1980) 1996. *The Literature of Terror: A History of Gothic Fictions from 1765 to the present day*. London and New York: Longman Group.

Rothberg, Michael 2009. *Multidirectional Memory: Remembering the Holocaust in the Age of Decolonization*. Stanford: Stanford University Press.

Sebald, W.G. (2001) 2002. *Austerlitz*. London: Penguin Books.

Shelley, Mary. (1812) 1994. *The Last Man*. Oxford: Oxford University Press.

Warwick, Alexandra. 1999. "Lost Cities: London's apocalypse." In *Spectral Readings: Towards a Gothic Geography*, edited by Glennis Byron and David Punter, 73–87. Basingstoke: Macmillan, 1999.

———. 2007. "Feeling Gothicky?" In *Gothic Studies* 9 (1): 5–15.

Žižek, Slavoj. (2010) 2011. *Living in the End Times*. London: Verso.

3 Not Now, Not Yet

Polytemporality and Fictions of the Iraq War[*]

Roger Luckhurst

Whenever I ask my students to think about defining the contemporary era through turning points, significant events or punctual moments, the conversation inevitably turns to New York on 11 September 2001. The event that day was shaped by the globalised flows of communication, transport, capital and geopolitical power, and so in turn the shorthand of "9/11" has become the spectacular condensation of those very dynamics. The cultural response to 9/11 was complex and multiform and the critical commentary on that response is now itself extensive and protean. In literature, statements were expected, and delivered, from the heavyweights: Don DeLillo, John Updike, Martin Amis, Art Spiegelman. Each new novel that addresses 9/11 is tested against whether it has approached making a definitive statement: witness the feverish superlatives that surrounded Joseph O'Neill's *Netherland* (2008) or Amy Waldman's *The Submission* (2011). Within ten years, we already have a clear idea of a certain canon of texts that enable familiar debates about the ethics and aesthetics of the representation of the catastrophe, as Kristiaan Versluys's study, *Out of the Blue,* exemplifies.

One cannot quite say the same for the Iraq war that followed as a far from logical consequence of 9/11 in 2003. There isn't yet the sense of defining literary texts emerging from the overlapping contexts of the war, the civil war or the occupation: symptomatically, perhaps, it isn't even clear how we should name, periodise or even characterise these events. When did it start? With the First Gulf War in 1991? Earlier? Is it separable from the war in Afghanistan, the longest military engagement in American history? And when did the Iraq War end? Has it ended? Barak Obama initially announced that the end of American military operations in Iraq would be in the summer of 2010, which was held by some to likely be as provisional as George Bush's notorious declaration of "Mission Accomplished" in 2003. So indeed it proved. Obama revised the date of final withdrawal to December 2011, a decision driven by the Iraqi government ending the exclusion of American military personnel from Iraqi law, thus necessitating withdrawal to avoid prosecutions (see Steele 2011, 28). Obama's inability to extricate himself from this post-war occupation has been the subject of much comment and much disappointment on the left (see Ali 2010 or Carter 2011).

My exploration in this chapter concerns how we might gauge the rela-tion of an unfolding contemporary war to cultural representations. Ethical criticism on the aesthetic representation of war is often pulled between the demand for witness and documentary record versus the call for indirec-tion, aporia or the foregrounding of the impossibility of representation of such traumatic violence. Modernist difficulty is often the favoured aesthetic mode. Yet this framework, generated in the main by critical commentary on the Holocaust, isn't necessarily helpful to transpose to contemporary events, where the urge to convey the hidden or suppressed consequences of violence in the most literal ways possible can have significant political impetus. For a certain time, Thomas Hirschhorn's art installation "The Incommensurable Banner" (2008), an 18-metre long collage of photo-graphs of ripped, opened, and mutilated bodies of Iraqi civilians, made up of images excluded from Western media representation of the war, has a powerful intervention to make. Yet contemporaneity needn't be marked by such literalist imperative, of course. In *War Cuts* (2004), the German artist Gerhard Richter juxtaposed news reportage from two days in the Iraq War alongside slices of his resolutely anti-representational abstract paintings, as if to pose this very question of adequacy and representation.

Yet can one even safely draw limits around what might be a "response" to the War on Terror in contemporary culture, less an event, perhaps, than a global network of confusing alliances and hidden complicities? And might not the violence of war disturb or disrupt the very notion of being con-temporary at all, shattering the illusory temporal order of a self-identical present? To pursue these reflections is to find, as I will shortly argue, that perhaps some of the most interesting cultural responses to the Iraq War in the West do not, in fact, ever directly mention the war.

Let us start, however, with a brief survey of the more obvious cultural engagements with Iraq. There is a very large body of cultural responses to the war, but it seems diffuse and isolated in specific aesthetic disciplines. It is striking that Stacey Peebles, in one of the very first book-length studies of cultural responses to the war traverses novels, short stories, blogs, poetry, comics and video games, a range of cultural responses that she argues are precisely marked by a desire to "transcend categorization," making them difficult to track and assess (Peebles 2011, 21). Peebles does not cover the reactions in the art world particularly in photography, but these are a good place to start, since there have been a plethora of photo-essay books, instal-lations, single shows, group exhibitions and even whole biennials (as in, for instance, the multi-site 2008 Brighton Photo Biennial, "Memory of Fire: The War of Images and the Images of War," curated by Julian Stallabrass). In his introduction to a *Photoworks* special issue, Stallabrass set out to situate Iraqi war images in a long comparative history to other wars and to confront the uncomfortable disjuncture of aesthetics and violence. After a long period of severe military restriction on press photography in the wake of Vietnam, where control of images effectively "virtualised" the First Gulf

War (a trajectory that has been outlined by Caroline Brothers), the strategy in Iraq was to allow freer movement but to embed the press corps. More images were available, therefore, but their complicities more difficult to grasp. The photographer Geert van Kesteren produced two books from his proximity to the Iraq invasion (*Why Mister Why?* and *Baghdad Calling*), while the video artists Wafaa Bilal and Harun Farocki have examined the strange effects that distant access to images of violence through networked computers can produce. Bilal invited internet users to "shoot an Iraqi" on an interactive website that directed a paintballing gun at the artist, while Farocki created the video piece, *Serious Games* in 2009, part of installation called *Images of War (At a Distance)*, which recorded the virtual training of American troops for Iraq through Pentagon video games played for training purposes in Twenty-Nine Palms in California. Photography became foregrounded early on in the aftermath of the invasion undoubtedly because of the uncontrollable circulation of digital images that emerged from the Abu Ghraib prison, which produced an intensely traumatic response and an outpouring of critical reflection on the life and death of images. Susan Sontag's initial reaction to the Abu Ghraib images, "What Have We Done?" (2004), almost seemed to challenge her established view that the ability for photographs to shock had progressively diminished since World War II. Stephen Eisenman, however, explored in *The Abu Ghraib Effect* why the images had failed to produce more outrage. Debates continued in *The Life and Death of Images* (Costello and Willsdon 2008) and Abu Ghraib continued to haunt both American and Iraqi artists, for example in the art of Ayad Alkadhi (see Safdar Ahmed 2011). At the 2013 Venice Biennale, the exhibition "Welcome to Iraq", curated by Jonathan Watkins, tried to reverse the angle and showcased eleven Iraqi-based artists commenting on the struggle to make art in the catastrophic aftermath of war.

The mainstream American media consensus on how the war was to be reported prompted a cluster of documentary works, including long-form narrative reports from embedded journalists such as Dexter Filkins, David Finkel, Patrick O'Donnell, and Sebastian Junger. Junger also reported from the forward American posts of Afghanistan in *War* (2010), and the filmed documentary he made with photographer Tim Hetherington about one such outpost, *Restrepo* (2010), was released in Britain as news emerged that Hetherington had been killed in the war to depose Colonel Gadaffi in Libya, prompting even more commentary on contemporary war reportage in age of asymmetric war. In a more dissident mode (away from the complex complicities of embedded reportage), Rajiv Chandrasekeran's reports from the Green Zone, *Imperial Life in the Emerald City* (2007), detailed the surreality and jaw-dropping incompetence of the American occupation, more than matching the weird world of Pynchon's post-war Zone in *Gravity's Rainbow*. Chandrasekeran's book was reinvented as Paul Greengrass's fiction film, *Green Zone* (2010), tied to a melodramatic conspiracy theory plot.

So powerful have these documentary interventions been that the writer Geoff Dyer has suggested that their quality and timeliness in bringing us stories and narratives about the Iraq War "has left the novel looking super-fluous" (Dyer 2010, 3). In cinema, there have been numerous documentary interventions that have unusually received major cinema releases, led by the leading documentarists of our time, Errol Morris with *Standard Operating Procedure* (2008) about Abu Ghraib and Nick Broomfield's *Battle for Haditha* (2008). Their release in cinemas says something about needing to circumvent the control of the circulation of contemporary images of foreign wars by the multinational conglomerates that control television news. Michael Moore's anti-Bush satire *Fahrenheit 911* (2004) even had a substantial role to play in the presidential election that year. But it is cinematic fictions about the war that have determined much of the iconography of contemporary asymmetric warfare: dusty checkpoints, handheld cameras, choppy edits, inscrutable Arabs, Humvees, IEDs, hooded prisoners, queasy torture scenes, vague liberal angst. Of the twenty-three Hollywood films focused directly on the Iraq War released between 2004 and 2009 (as counted by Barker), many were structured around a model of a returned veteran suffering post-traumatic stress, building towards the narrative revelation of the repressed traumatic event from the war. By 2009, *The Messenger* was close to reducing this form to rote. Even so, a lot of the commentary on the cinema of the Iraq War has discussed its aesthetic failures and its striking inability to attract audiences or gain traction as significant cultural statements in the same way as *Deerhunter* or *Apocalypse Now* defined a certain iconography of Vietnam, as the film critic Ali Jafaar has argued. Ben Fountain's novel, *Billy Lynn's Long Halftime Walk* satirizes Hollywood's attempts to turn traumatic Iraq war experience into Hollywood conventions. However, the story of the poor aesthetic or financial returns that has attended these films, almost from the beginning, has prompted Martin Barker to suggest that the *rhetoric* of failure has been a way of neutralising the challenge of many of the films that work '*to undermine presumed ways of understanding the war* and to *provoke disquiet*' (Barker 2011, 118). The relative success of Kathryn Bigelow's *The Hurt Locker* (2008) was premised on its evasion of overt political statement by a pathologically narrow dramatic focus, but even an Oscar for its director failed to lift its minimal distribution deals and meagre returns. After an enraged right-wing press hounded Brian de Palma for his dislocated denunciation of the war in *Redacted* (2007), distributors in America simply refused to show Bigelow's film. Paul Haggis's *In the Valley of Elah* (2007) was similarly abused for its "liberal" desecration of the American flag in the closing scene, after a father discovers his son was involved in the torture and murder of Iraqi prisoners and raises the flag upside-down, a traditional distress signal.

Meanwhile, there is a significant archive of war poetry and new forms of witness, often exploiting new technologies of writing in online journals and blogs that sidestep the publishing industry, largely ignored by main-stream literary commentators. Brian Turner's *Here, Bullet,* published in 2005 following his year's deployment in Iraq, was followed by *Phantom Noise,* which was shortlisted for the prestigious T. S. Eliot prize in Britain in 2010. Turner was one the strongest poets to emerge from the *Voices in Wartime* anthology, but the internet is bursting with thousands of pages of self-authored material and anti-war poetry collections. Yet, it must be said, poetry has a thinning presence in the contemporary public sphere, something the poet Simon Armitage addressed in his introduction to the script of his poetry-film, *The Not Dead* (2008).

The war coincided with the explosion of Web 2.0, the internet as a vehicle for self-authored publication. This not only increased the availability of poetry, but also new expressive forms such as blogs by soldiers. By 2013, the aggregating website milblogging.com was tracking nearly four thousand separate blogs by military personnel. Because of the immediacy of the form and an ability to evade official military censorship, early responses to the American invasion and the difficult aftermath through these new modes of record were crucial (the uploading of images and videos to sites like YouTube were a constant source of anxiety and scandal for the authorities trying to control information from the frontline). An anthology of American blog narratives was collected by Matthew Curier Burden in 2006, but Iraqi voices that blogged from inside occupied Iraq, such as the celebrated reports of "Salam Pax" and the sardonic female blogger "Riverbend" were also crucial to round out war reportage.

For a defeated anti-war left, whose support peaked with the global anti-war marches of 15 February 2003 (marches that appeared only to galvanise the messianism of Bush and Blair), an odd celebrity has since accrued to a number of intellectual dissidents writing out a critical theory of war. Commentaries tumbled out of the Communist Lacanian Slavoj Zizek, from *Iraq: The Borrowed Kettle* (2004) onwards. The more patient reflections of Judith Butler on violence and the brutal exclusion of wartime enemies from what is human or grievable explicitly addressed Iraq in *Frames of War* (2009). Quieter still was the philosopher Adriana Cavarero's coinage of the term *horrorism* for the unique depredations of contemporary violence that have extended beyond its analogue, terrorism. Cavarero's *Horrorism* is a philosophical investigation studded with moments of reportage from the occupation and civil war in Iraq, with a particular focus on the destruction of the defenceless. With the new word *horrorism,* Cavarero says she wants to point to "something new, different, recent" in modern war zones, using a coinage that "helps us to see that a certain model of horror is indispensable for understanding our present" (Cavarero 2009, 29).

The conventional account is that there is a traumatic time lag between war and its novelistic assessment, as if the immediacy of war experience is irrecoverable. But there has just as often been a literature of instant reaction to war. Yet it is hard to think of prose fiction about Iraq that has had the kind of cultural impact that might rival something like Norman Mailer's *Why Are We in Vietnam?* (1967) or Tim O'Brien's *If I Should Die in a Combat Zone* (1969), which emerged at the height of the engagement. In the early years after 2003, polemical novels about Iraq were published by small presses. Tony Christini's *Homefront* (2006), for instance, is a tendentious diatribe that sacrifices nuance and ambiguity for unclear ends, since documentary forms can occupy this ground more authoritatively. The elliptical approach of Benjamin Percy's short story collection *Refresh, Refresh* (2007), with its evasive and inarticulate Iraq veterans, lost siblings and under-determined grief-stricken back-stories seem more forceful. It is perhaps symptomatic that Percy's stories morphed rapidly into a film script and then again into a comic by Danica Novagodoroff within a year of publication. Percy's decision to use the genre conventions of horror fiction in his novel *Blood Red Moon* (2013) also seems significant, since the book uses a grinding, endless struggle between humans and werewolves to narrate displaced versions of the globe's contemporary wars. Thriller forms have incorporated Iraq contexts, given their interest in global networks and interconnections (as in Lee Child's *Nothing to Lose*), but then conspiracy is, as David Pascoe has suggested, the predominant form of war writing since the height of the Cold War. Crime books have rapidly consolidated the cliché of the returning soldier, even breathtakingly consigning Iraq to a red herring, exploiting our stereotype of veterans with Post-Traumatic Stress Disorder, as in Minette Walters's *The Chameleon's Shadow* (2007). While several American writers tried to occupy the mind of the 9/11 terrorists, it has been left to others to explore the radicalisation of Iraqi populations in the phase of revolt against occupation. The second person address of Mohsin Hamid's *The Reluctant Fundamentalist* (2007) is meant to unsettle and force a recognition of Western complicity in the murderous intensifications of the "war on terror."

It is direct experience of war that seems to guarantee a certain authenticity, however. Yasmina Khadra's *The Sirens of Baghdad* (2008) is directly located in the aftermath of the American invasion of Iraq, an excoriating study of the making of an insurgent. The novel was written pseudonymously by the Algerian army officer Mohammed Moulessehoul and translated from the French in 2008. Khadra sold well in Europe. American reviewers immediately embraced Kevin Powers's *The Yellow Birds* (2012) because his experience as a machine gunner in Mosul in Iraq in 2004–05 was expressed through a finely honed MFA literary style that suitably transfigured the traumatic details of the tour into polished *aperçu*. Any chance of grasping the Iraqi experience of the war has had to rely on translations from specialist academic presses, as in the *Contemporary Iraqi Fiction* collection put together for the American University in Cairo by Shakir Mustafa. The journal *Words without Borders* published a special issue, 'Iraq, Ten Years Later' in April 2013, featuring translated Iraqi fiction by Hassim Blasim, Najem Wali, Ali Bader, and others, with a critical survey of the current Iraqi writing scenee by Yasmeen Hanoosh (see http://word-swithoutborders.org/issue/ april-2013). Sinan Antoon, an Iraqi-American writing in exile, has published extraordinary work, and although *I'jaam: An Iraqi Rhapsody* is at the very limits of translatability from Arabic to English, the desperate vision of post-conflict Iraq in *The Corpse Washer* conforms more conventionally to the Western novel form and may get more notice as a consequence.

Novelistic commentary in England was meant to come from Ian McEwan's *Saturday* (2005), which arrived weighted with expectation, but which proved weakly contrived. *Saturday* is set on the day of international protests against the war in February 2003, two months before the campaign of Shock and Awe began. All the contrivances come in trying to strain the global through the local, following the professional and domestic travails of the neurosurgeon narrator Henry Perowne in his privileged London life against geopolitical events, using the model of Virginia Woolf's *Mrs Dalloway* (1925). Where Woolf's reflections on the legacy of war are all the more devastating for their fleet obliqueness, McEwan's heavy-handed plotting reaches its peak of awkwardness when literature, in the guise of Matthew Arnold's "Dover Beach," is located as an irenic *deus ex machina*, a humanistic balm that magically diffuses the threat of violence, but which so incompetently handled rather only works to reinforce the marginality of literary discourse. A painful self-consciousness about whether literature is appropriate at all for capturing the post-9/11 world is what drives James Meek's *We Are Now Beginning Our Descent* (2008), featuring a traumatised war reporter, peddling a commercial and conservative blockbuster on the war in Afghanistan in 2001–02 instead of the difficult art he senses the situation requires but which he has no courage to write. The novel opens with an Afghan scene written in taut thriller prose, but the reader is disarmed several pages in by the passage suddenly breaking off into a punishing critique of its clichés by the central character. Adam Kellas is only one of many war reporters who recur in contemporary fiction, perhaps marking a sense of crisis about the ethics of fictional representations of the violence of modern war (novels that include Michael Ignatieff's *Charlie Johnson in the Flames* (2003), Pat Barker's *Double Vision* (2004), or Andrew Miller's *The Optimists* (2006)). This sense of crisis tends to reinforce the position of David Shields in *Reality Hunger*, when he argues that the art of the novel has been broken open by the need to incorporate "larger and larger chunks of 'reality'" into prose-forms, and that consequently the "most compelling energies seem directed at nonfiction" (Shields 2010, 3 and 26). One might even suspect that it is the "problem" of Iraq that is one of the main elements behind the Shields thesis.

The year 2012 seemed to mark a threshold when novels about Iraq finally began to appear in substantial numbers from mainstream American presses (see Harris 2013). Aside from Powers's *The Yellow Birds*, there was the satirical tone of David Abrams's *Fobbit* (2012) or Ben Fountain's

Billy Lynn's Long Half Time Walk (2012), both of which settled the Iraq War into the frameworks provided from earlier wars by the absurdism of Thomas Pynchon or Joseph Heller. A serious fervour to expose untracked traumatic experience still drives Helen Benedict's *Sand Queen* (2011), however, one of the first fictional recognitions that the Iraq War was the first war in which substantial numbers of American women experienced forward operations.

There are obvious reasons why this muted or diffused reaction to the war might predominate. 9/11 was a single punctual event that, for all the language of singularity and unprecedentedness, was strikingly easy to narrate within the paradigm of trauma. The framework of trauma had already been transferred from psychology to culture in the 1990s: 9/11 conformed to diagnostic notions of the event outside the bounds of normal experience that produced not just large scale death and wounding, but an uncontainable contagion of traumatic secondary witness through its calculated media spectacle. This formulation of what I have elsewhere called the trauma paradigm, relies as much on the diagnostic language of the American Psychiatric Association as more cultural theories of trauma. The punctual event of the attack on the World Trade Center was *intended* to produce a distinctive aftermath, and this state of after-shock was eminently readable in the discourse of post-traumatic reaction, at individual, community and national levels.

In contrast to this highly readable event, the Iraq War exists in an odd stage of incompletion, at once a war, a civil war, and a post-war occupation, an intervention that began as an ostensibly symmetrical engagement between armies that mutated into asymmetrical guerrilla warfare, insurgency, and what now looks like the classic violent aftermath following colonial withdrawal, with destructive indigenous factional fighting. The politics of the war remains intensely divisive, and, for the American public, the sympathies deeply confused. Fighting a war of vengeance in the rhetoric of liberation, ultimately premised on either fabricated or manipulated intelligence, American soldiers are portrayed at once as victims of a merciless military-industrial complex or a dastardly Iraqi resistance, but also simultaneously as ignorant 'grunts' and perpetrators of uncountable deaths of non-combatant civilians. The military provided their own traumatic iconography of the logic of occupation in the images of Abu Ghraib. The circulation of these images, precisely because they horrifyingly condense many of the disastrous aspects of an unprepared post-invasion strategy, has renewed the punctual language of trauma. Much American cultural commentary on this curious phase of post-war war in Iraq has been routed through the shock that the images from Abu Ghraib produced. Elsewhere, this military, political and ethical quagmire, without foreseeable end, has not made for easy narrative contours or crystallising representations. In many ways, Richard House's monumental novel in four books, *The Kills* (2013) is emblematic: the Iraq War is the penumbra of this sprawling account of the corruption and confusion in the wake of the invasion, the novel featuring only confused civilian contractors, mercenaries, corrupt officials and the mysterious emptiness of spaces ravaged by a war that is left entirely undepicted. This is what Ross Chambers once called *aftermath culture,* defined "by a strange nexus of denial and acknowledgement of the traumatic" that leaves a haunting trace everywhere (2004, xxi). Perhaps another lesson from the American engagement in Vietnam is that wars need a definitive end, however ignominious, before there can be any sustained cultural reflection.

Perhaps, though, in those first years we were always looking in the wrong places for cultural reflections on the Iraq War. The thesis I want to pursue here is a relatively simple one: the resistance to narrative or representation of the contemporary war means that cultural narratives about it are often displaced or filtered through the iconography of prior wars. A slightly different but related thesis might be that this refraction is the *only* way of grasping war in its contemporaneity.

This is not a new insight. At the beginning of "The Eighteenth Brumaire of Louis Bonaparte," Karl Marx famously suggests that "all the great events and characters of world history occur . . . twice . . . : the first time as tragedy, the second as farce." Leaders who pronounce their revolutionary newness, Marx said, "conjure up the spirits of the past to help them; they borrow their names, slogans and costumes so as to stage the new world-historical scene in this venerable disguise and borrowed language" (Marx 1981, 146). Thus begins Marx's searing critique of Napoleon's nephew in 1848 dressing up as the great general from 1793 who in turn draped his putsch against the Directory in the clothes of the Roman emperors. Yet this insight, that the newness of the contemporary revolution actually prompts a kind of multi-temporal overlay of different revolutionary times, has become intrinsic to recent theories of the contemporary as such. Giorgio Agamben in his short essay, "What is the Contemporary?", defines it through its necessary untimeliness: "Only he who perceives the indices and signatures of the archaic in the most modern and recent can be contemporary" (Agamben 2009, 40). It is "through this disconnection and this anachronism," Agamben continues, that those rare figures who become significant contemporaries "are more capable than others of perceiving and grasping their own time" (Agamben 2009, 50). A similar stance was explored more robustly by Steven Connor, who suggested that the contemporary is best grasped as the contemporal, a mixing of times together, which is understood, following the work of Michel Serres and Bruno Latour, as the intrinsic condition of the "impossible" present. For Connor, "in contemporality, the thread of one duration is pulled constantly through the loop formed by another, one temporality is strained through another's mesh" (Connor 1999, 31). There *is* no time like the present, in this reading of the contemporary, only a constant interplay or percolation of one through the other of the old and the new.

If the representation of the traumatic consequences of war produces the same contemporal effect, that may also be because this is reinforced by the disturbances in temporality that have long been understood to attend traumatic events. Central to cultural-critical trauma theory, of course, has been Freud's emphasis on *Nachträglichkeit,* translated as deferred action or afterwardsness, where the first moment of trauma is grasped through a later, secondary moment that only retroactively generates traumatic force in an odd time after time (the most influential use of this Freudian concept has been in the work of Cathy Caruth). In trauma, as it is understood in this account, the contemporary is ghosted or haunted by an insistent past that intrudes on, overlays and redetermines the present. More recently, and in an important extension of trauma theory, Michael Rothberg has argued for "multidirectional memory," which places this kind of polytemporal memory at its heart. Rothberg resists models of competitive memory, where, most notoriously, some Holocaust memorialists declare its exclusivity and exceptionality over all other acts of systematic murder or genocide in a zero-sum game. In place of a struggle for pre-eminence between events, Rothberg offers a series of innovative readings in which the traumatic events of one time and place are read through the filter of another. So, for instance, French writers in the early 1960s insistently overlay the atrocities of the Algerian War on the murderous logic of the concentration camps. Denunciations of imperial atrocities after 1945 repeatedly invoke prior models of Nazi atrocity. Yet this does not happen in one direction only, for the outrages in Algeria in the early 60s oddly allow for retroactive articulations of the experience of the camps, as if for the first time, as if the meaning of the Nazi genocide can actually only emerge through the very act of comparison to contemporary atrocity. For Rothberg, the multidirectionality of memory is meant "to draw attention to the dynamic transfers that take place between diverse places and times during the act of remembrance." It acknowledges, he continues, "how remembrance both cuts across and binds together diverse spatial, temporal, and cultural sites" (Rothberg 2009, 11). For my purposes, then, the intrinsic multi-temporality of traumatic memory translates into something like this structural law: one war time will always be seen through the lens of another.

I want to put this theory to the test by looking at three very different texts that emerged in the same short period of time: Denis Johnson's historical novel about the Vietnam War *Tree of Smoke,* published in 2008 and winner of a National Book Award, Guillermo Del Toro's film *Pan's Labyrinth* from 2006, a hybrid mix of fantasy and history set at the very end of the Spanish Civil War, and Kathleen Ann Goonan's award-winning science fiction novel, *In War Times,* from 2007, the plot of which is generated by directly addressing the polytemporal nature of traumatic war experience. This eclectic choice is deliberate: just as I have always considered it vital to grasp the trauma paradigm across different cultural forms, and across high and low culture, so it is important to trace the diverse generic sites where the question of Iraq might be seen to erupt in contemporary culture. None of my chosen texts in fact directly mention the war in Iraq, and all variously stretch to evoke historical authenticity of prior epochs. At the same time, they seem to speak of little else but our contemporal war.

Denis Johnson's sprawling epic, *Tree of Smoke,* runs from 1963 to 1969, with a short coda from 1983. This time-frame is foregrounded in the structuring of the novel, intertitles insistently situating the reader in this historical sequence. Johnson attempts to provide a genealogy of the escalation of the American intervention in South-East Asia, opening with a long section in Indo-China as French power crumbles and an already paranoid American secret service engineers the assassination of an innocent missionary. The novel takes nearly three hundred pages to arrive at an evocation of the battle experience of American soldiers on the front-line in Vietnam, in which combat time dilates and contracts disarmingly and any sense of direct experience is radically dissociated from the consciousness of the soldier. There is no battle experience as such, and in this, Johnson conforms rigorously to trauma theory. We are left only with the panic of a retreat, the idiocy of soldiers firing flares at their own commanding officers, and a confusion of a war that cannot even be reconstructed subsequently from the reports and overviews published in *Time* or *Newsweek*. One character "found it all written down there, yet these events seemed improbable, fictitious. In six or seven months the homeland from which he was exiled had sunk in the ocean of its future history" (Johnson 2008, 329).

There is a certain studied banality about Johnson's meandering plot and dialogue, and a desperate familiarity to his scenes of dissolution, corruption and even elaborate torture of suspected Viet Cong guerrillas. This has prompted some vigorous criticism, including B. R. Myers's impressive denunciation in *The Atlantic Monthly* of *Tree of Smoke* as offering only "a renewed sense of the decline of American literary standards" (Myers 2007). Myers takes Johnson's style apart at the level of the sentence. But this awkwardness and banality is surely deliberate: the book has to be read through the litany of cultural clichés about Vietnam War experience, and the language of deadened effect, dissociation, and the psychological trauma of war were precisely forged by military and counter-cultural psychiatry in the crucible of Vietnam. As is now well documented, post-traumatic stress disorder entered the official diagnostic manual of the American Psychiatric Association in 1980 as an extension and generalisation of combat stress that advocates working with Vietnam veterans had been campaigning to be recognised as a compensable mental illness. Johnson thus exhaustively rehearses what we already know in deadened post-traumatic prose.

Yet what most of the early reviewers of the book failed to engage with was another genealogical project at the heart of the novel, in which Johnson explores the origins of the Iraq War. The "Tree of Smoke" of the title is a nascent secret intelligence database, a box of index cards obsessively cataloguing the world, which has been constructed by the enigmatic Colonel

Sands, a maverick on the edges of the military and the OSS as it mutates into the CIA. Sands's tireless work of destabilisation sees him at the origin of many regional conflicts in Asia. The mysterious colonel, who in a rather Pynchonesque manner simply disappears into the matrix of his own clandestine smokescreens half way through the book, stays out in the field because his superiors do not appreciate his highly critical document on the dangers of what he calls "command influence" on intelligence gathering. This is exactly the point that Iraq obtrudes through Vietnam. "Simply put," Sands says in his paper, "the need to examine the veracity of sources yields to the pressures of process. The result is cross-contamination: data from human sources, notoriously undependable, become the support for doubtful interpretations of documentary sources, and these interpretations come to be seen as shedding light, in turn, on data from human sources" (Johnson 2008, 251). Banging the point home, the document continues: "the interpretive process, we remember, is always subject to appropriation and enlistment in the service of policy. Cross-contamination renders data vague, malleable, and eventually useless as anything but an ingredient of internal bureaucratic and political chemistries" (Johnson 2008, 251). In a hand-written addition to his unfinished paper (perhaps an echo of Kurtz in Conrad's *Heart of Darkness*), which lies disregarded and unpublished, having ruined his prospects, Sands adds: "the next step is for career-minded power-mad cynical jaded bureaucrats to use intelligence to influence policy" (Johnson 2008, 254). At this point, *The Tree of Smoke* becomes less an historical evocation of the conditions of the Vietnam War and more an ambitious long-term analysis of how the American intelligence agencies so catastrophically failed to gather trustworthy field intelligence on Iraq, and let policy be dictated by motivated Iraqi exiles and the neo-Conservative ideologues around George W. Bush. As soon as this document appears in the novel, it demands to be read polytemporally, the novel focalising one war through another. Johnson's genealogy has impressive "multidirectional" polytemporal effects. The necessity of victory in Iraq for America was repeatedly framed in opposition to the defeat in Vietnam, a laying to rest of a humiliating failure, hailing a definitive era of succession from sapping colonial engagement. Johnson's novel instead suggests the continuity of a certain illogic in which the American military-industrial complex generates its own murderous address to others who are often built up from minor threats to become monstrous phantasms of its own paranoid intelligence bureaucracy. The insights are generated by having one war time superimposed upon another, seeing Iraq through Vietnam but also Vietnam through Iraq.

In a heightened historiographic-realist mode, this aspect of Johnson's project does not seem to me to be difficult to discern, although it merited virtually no comment in reviews. In contrast, Guillermo Del Toro's film, *Pan's Labyrinth*, is more challenging to categorise and read, at once fairy-tale, quest romance, portal fantasy and historical war film. The film is set in 1944 in the strange period after the Spanish Civil War when the war against Franco's fascist regime by surviving rebels had not quite ended. It weaves together the last stand of a group of partisans against Fascist troops with a fairy tale quest in which the daughter of the house, Ofelia, is set tasks by a woodland faun to repair a wounded kingdom that exists in a different order of reality, accessed by various portals around the ancient farmhouse, most importantly through a pagan labyrinth. *Pan's Labyrinth,* which follows Del Toro's other Spanish war film in Gothic mode, *The Devil's Backbone,* could be situated as a text emerging after a lengthy period of the suppression of collective memory about the Civil War in Spain. In a study of cultural production in Spain after the Transition from Fascist rule in the late 1970s, Eloy Merino and Rosi Song argue that contemporary Spanish culture is now full of displaced echoes of the war: "It is the prohibited nature of these traces that gives them their ghostly quality, so that they can function undetected, unnoticed, and banned from normal communication" (2005, 17). It would be possible to read Del Toro's use of fairy tale tropes as an evasion of this form of silencing, the film responding, as Paul Julian Smith has commented, to "the frequent accusation that democratic Spain has turned its back on traumatic history" (2007, 8). This sounds like a familiar national variation on the belatedness of dealing with the traumas of a collective memory. But towards the end of *Pan's Labyrinth,* there are two graphic scenes of torture which one inevitably comes to watch through the filter of the American military occupation of Iraq. Rather like Ken Loach's *The Wind that Shakes the Barley* (2006), a film about the Irish Civil War in the 1920s, the scenes of torture that invade so rudely into the art-house register feel like intrusions of very contemporary disorders of violence. This is something I've argued in more detail elsewhere (Luckhurst 2010). In this decade, the scene of torture inevitably invokes Abu Ghraib and Guantanamo, the elaborate networks and euphemisms of America's process of extraordinary rendition.

In this context, Del Toro's scenes carry a deliberate polytemporal freight. The film works, in Rothberg's terms, "to draw attention to the dynamic transfers that take place between diverse places and times during the act of remembrance" (2009, 11). In the Spanish context, does the unfinishedness of Civil War trauma allow us to read further into *Pan's Labyrinth* multidirectionally, seeing it as a refracted engagement with the pro-invasion stance of the conservative government under José Maria Aznar, one of the few international members of Bush's "Coalition of the Willing"? Aznar lost the Spanish election in 2004 only ten days after the Madrid train bombings that killed 191 people, which Aznar had initially blamed on Basque separatists, rather than as a consequence of his commitment to the Iraq invasion. By May 2004, all Spanish troops had been withdrawn from Iraq by the new government. Yet the legacies of violence, *Pan's Labyrinth* suggests, are not so easily managed, but well up and overwhelm temporal causation.

The complexity of one's reaction to the film is reinforced by the fluidity of movement between the registers of history and fantasy. The fantastic is often conceived of as a strictly demarcated space of impossibility, reached through portal or distinct threshold dividing real and unreal. In *Pan's Labyrinth,* the camera movements continually sweep across levels of adult war and childhood fantasy without disruption, each one bleeding continually into the other, confusing confident categorisation. Pan, half-goat, half-god, both earthy and divine, encourages this hybridity. Pan or Faunus might be associated with the pastoral, but he is also a trickster, famed for inducing pan-ic, particularly in soldiers. Myth feeds specific historical representation.

In her recent critical study, *Translating Time,* Bliss Cua Lim has argued that fantastic cinema is intrinsically concerned with what she calls, following Bergson, "nonidentical temporality." Fantastic cinema, she says, always "opens to more than one time" (Lim 2009, 8). Lim focuses on the Japanese and Korean supernatural films of the 2000s, ones like *The Ring* and *The Grudge* that went through rapid cycles of sequels and American remakes in very short order. The vengeful spectres that menace the living in these genre texts are for Lim shards of one time slicing through another, instances of temporal anomaly that fracture the unitary time of modernity. There is a postcolonial critique here too, in that local spiritual or supernatural beliefs resist the temporal containment of difference into the singular time of secular disenchantment. By overlaying these critiques, Lim argues that the fantastic mode "can disclose a starting point for temporal critique, one that is enmeshed in the very idiom of homogeneous time yet strains against it, producing a quality of uncanniness" (Lim 2009, 26). Although Lim only focuses on the ghost film, and thus the obtrusion of the recalcitrant past into the contemporary, very familiar to us from Jacques Derrida's account of the polytemporal ghost in *Spectres of Marx*, a wider definition of the fantastic would incorporate Del Toro's temporal disjuncture produced when he forces historical and mythic time to occupy the same liminal space in *Pan's Labyrinth*. We might also include in the intrinsic multi-temporality of the fantastic the science fiction genre.

The imagination of future wars often models conflict on past wars, and is thus an allegorical form that we must consider polytemporal. But it is also a genre with a slightly more disturbing relationship to warfare in the twentieth century than might at first be expected. Hawkish American science fiction dreamed the ultimate atomic weapon into being long before the Manhattan Project and subsequently rehearsed outcomes for the Cold War for decades on a parallel track to the enthusiasm of actual Pentagon war gamers. It was a group of right-wing science fiction writers around Robert Heinlein that persuaded Ronald Reagan to fund Project High Frontier, which became the Strategic Defence Initiative or 'Star Wars' in the 1980s, an extraordinary story that has been investigated by H. Bruce Franklin. Meanwhile, a sly liberal tradition has consistently offered a critique of the technological fetishism and the logic of spectacle at the heart of the Cold War, most consistently in the work of Philip Dick or Frederick Pohl. For a linkage of war, trauma and temporal disturbance in the genre, one needs only gesture towards three central figures in post-Second World War science fiction. In Kurt Vonnegut's *Slaughterhouse 5* (1969) Billy Pilgrim is scattered through time by the very force of the bombs rained down on Dresden in February 1945. J. G. Ballard's apocalyptic visions of slowing and thickening of temporality in seminal texts like "The Voices of Time" (1960) or *The Crystal World* (1966) lend themselves to being read as translations of his war experience of internment in a civilian camp in Shanghai at the end of the Second World War. These texts, as the critic Paul Crosthwaite has suggested, are "a secret code of pain and memory" that compulsively repeat war trauma, although the temporal disturbances involved make the situation more complex than that (2009). Meanwhile, one of the best novels about the Vietnam War remains Joe Haldeman's *The Forever War* (1974), in which veterans of a future war suffer traumatic dissociation from Earth population because the "time dilation" of their interstellar travel to battlefields leaves them centuries older than their long-dead contemporaries and utterly alienated from the bewildering cultural shifts on the home front to which they have no lived connection. Haldeman fought in Vietnam and had tried writing a realist novel about a returning Vietnam veteran before hitting on the device of science-fictional estrangement (that realist novel, *1968*, was published later). Haldeman's science fiction reiterates war experience as, quintessentially, a disadjustment in time.

For my final instance of a polytemporal text on the Iraq War, though, I want to turn to Kathleen Ann Goonan's *In War Times*, a book which won the John W. Campbell Award for best novel in 2007. Goonan has written an acclaimed quartet of novels about nanotechnology set in a radically other far future. Her interest in future technological transformations of the human ironically, given her liberal politics, led to an invitation to address Pentagon researchers on future warfare during Bush's time in office. *In War Times* is a very different novel. Much of the book is set in the Second World War and is actually woven around the real war-time diary of her father Thomas Goonan, who worked as an engineer and followed close behind the front-line of the American liberation of Europe in 1944 and 1945. He ended up repairing the telephone networks the allies had often only days before knocked out in their advance on German positions.

However, around this authentic war-time document, Goonan weaves a fantastic tale. One of the brilliant Jewish refugee scientists from Europe attached to the American war effort is conducting her own secret project within the interstices of the Manhattan Project. While the military-industrial complex is being formed around the race to produce the atom bomb, Eliani Hadntz is instead directing her scientific researches towards an irenic end. Through an innovative combination of biological, technological and psychological research, scrawled in weird, half-finished scientific papers far in advance of scientific knowledge in the 1940s, Hadntz seems to have designed a quantum computer, with the aim of shifting human consciousness from its hardwired tendency towards violence. Hadntz has the plans but no means to build her prototype: she leaves these mazy, barely comprehensible documents with Sam Dance, the hapless engineer protagonist of the novel. He fumbles with the schematics while Hadntz enigmatically appears and disappears, taking him on impossible journeys into concentration camps and the underground workshops of Camp Dora where the Nazis are building their "revenge weapons," all in order to urge Dance to construct the device. Dance is even given access to the surveillance aircraft that follows the Enola Gay as it deploys the atom bomb above Hiroshima. Dance thus bears the weight of the war's atrocities, aside from his personal losses, a brother killed in the attack on Pearl Harbour.

The device he and his friend Wink construct seems unsuccessful at first, a dud. Yet it seems to have the capacity to self-evolve and it transpires that they have built a peculiar sort of time machine. The device actualises different temporal branches flowing from every instant, so that whilst Sam returns from Europe to our post-war world, his friend Wink ends up in a parallel post-war. Glimpsed possibilities from alternative pasts multiply: Sam tunes one of the versions of the device like a radio, only to conjure up his dead brother, as if he had lived beyond Pearl Harbour. Hadntz's own war trauma has propelled her research: she has lost much of her family in the concentration camps. These brachiations of time thus teem with loss, but also promise glimpses of plenitude if one can learn how to navigate through the multiplying time streams.

Goonan figures the quantum brachiation of these myriad times as if it were a bebop jazz improvisation; Sam and Wink see Charlie Parker play in New York and Parker's ability to hear different time signatures, or transfer between chromatic scales in a new way, is offered as a form of quantum consciousness. Their research is paralleled in their own manic experiments playing bebop jazz. This in fact has intrinsic connections to the Cold War, since jazz became part of the cultural front developed by American political strategists to promote a vision of a free, liberal capitalist culture, as cultural histories like that of Lisa Davenport have explored. Thus the polytemporal post-wars that proliferate out of the Hadntz Device are an improvised music of the spheres. Interestingly, Steven Connor also uses the metaphor of music to convey his notion of the contemporal. "The scoring of time constituted by one temporality," Connor says, "is played out on temporal instruments for which it may never have been intended, but which give it its music precisely in the way they change its metre and phrasings, and remix its elements" (2009, 31).

Goonan's novel is driven by an aching sense of the losses not just of the Second World War, but of the trajectory American politics took after 1945. There are some rare moments of nexus when Sam and Wink's alternate realities allow them to meet. In an ultimately rather didactic way, Wink's post-war world is presented as a utopian alternative to ours, a place where the Cold War did not develop, where peace overcomes conflict, where technological advances were shared between Russia and America, and a planetary politics emerged that fosters irenic technological progress, including meaningful space exploration. Wink appears at these junctures to warn that the endlessly proliferating temporalities flowing from the Hadntz device are nevertheless all threatened by the violence and destructiveness of *our* time, *our* zeal for weapons of mass destruction. The truly pivotal moment for post-war American history, it transpires, is the assassination of John Kennedy. In the last section of the book, Sam and his family travel back in time to Dallas in an attempt to avert the conspiracy and thus rescue *all* of the polytemporal possibilities threatened by this nexus.

In 2011, Goonan produced a sequel, *This Shared Dream,* which focused on the lives of the three Dance children, who suffer odd temporal slippages between multiple post-war Americas. For them, the base-line reality branches off from the family's successful frustration of the assassination of Kennedy, a world without the Vietnam War, without the social divisiveness of the 1960s and full of substantial liberal improvements, yet also haunted by traumatic personal losses, since both their parents effectively disappear into other timestreams to continue the war at this brachiation point. The Second World War is indeed unending for their parents and Eliani Hadntz, and their plans to disseminate the Device into the very brain development of the global population, ensuring a *biological* end to hard-wired hierarchical and aggressive behaviour, seems to win out against an ancient Nazi conspiracy by the end of the novel. Yet this slightly more benign alternative present is also located in 1991, as if the legacy of the Second World War can be beaten off but the characters can't yet see that the First Gulf War in our timestream will tilt against irenic utopianism and begin a different geopolitical trajectory to the American addiction to war.

It is not difficult to read another genealogy of the Iraq War in Goonan's fiction, although it may be that it is an entirely readerly effect.[1] I would read the contemporary war as the product of what has been called "The World War II Regime," defined by Andy Pickering as the way "social, technical, material, and conceptual developments, the civilian/scientific and military communities redefined themselves around endpoints that were interactively stabilised against and reciprocally dependent on each other" (1995, 18). The intervention in Iraq was partly generated from that still-dominant matrix of American military-industrial capitalism and the geopolitics that has grown from it. Goonan's cry for peace risks replacing politics with a reductive biological solution to war and her futures cannot see a way around a model of American capitalism that merely needs its logic of domination reformed, as if its warring tendencies were accidental rather than structural to its development. Yet despite its (irenic) reformism, *In War Times* delivers from the plural in its title onwards another instance in which we can grasp the polytemporality of our apprehension of the Iraq War.

This analysis could be extended across a further array of texts, from Rachel Seiffert's novel *Afterwards* (2007), say, which offers an oblique encounter of traumatised veterans from Britain's colonial wars in 1950s Kenya and 1970s Ireland, to the sophisticated musings on Iraq that were consistently foregrounded in the revamped SF series, *Battlestar Galactica* (2004–09), in which time looped in odd and unpredictable ways. A TV series like *Fringe* (2008–13) also figured multiple times and alternate universes for contemporary America, in a story marked by personal and national losses insistently figured through the traumatic absence of the World Trade Center towers and the consequences of the war on terror America continues to fight. Whatever the intrinsic value of these instances, they all illustrate my case that contemporaries grasp the contemporaneity of war through the refractions of the polytemporal. It might be that in the first years after the 2003 invasion the less cultural work appears to address the Iraq War, the more it actually does.

*This is an updated version of an essay that first appeared as "In War Times: Fictionalizing Iraq." Originally published in *Contemporary Literature* 53.4 (Winter 2012): 713–737. © 2012 by the Board of Regents of the University of Wisconsin System. Reproduced courtesy of The University of Wisconsin Press.

I benefited immensely from anonymous readers, the input of editors of the special issue "Fiction Since 2000", David James and Andrzej Gasiorek, and from the editorial work of Mary Mekemson. I am grateful for the permission to update and revise this essay. The first impetus to reflect on this came from Monica Calvo and Marita Nadal: thanks to both.

NOTES

1. I asked Kathleen Ann Goonan about whether she had wanted to embed a connection between World War Two and the Iraq War in the Dance books. Her intention at least neatly resists my thesis: "There is no specificity about Iraq in *This Shared Dream*. World War Two, including the Cold War, is the central war narrative in both books, and it amply portrays us at our worst. It is my metaphor for all war. The roots of the Iraq War are, of course, wildly different than those of the U.S.'s involvement in World War Two; so much so that it seems to be the face of a newer monster" (email 25 October 2011).

WORKS CITED

Abrams, David. 2013. *Fobbit*. London: Harvill Secker.
Agamben, Giorgio. 2009. "What is the Contemporary?" In *What is an Apparatus? And Other Essays*, translated by D. Kishik and S. Pedatella, 39–54. Stanford: Stanford University Press.
Ahmed, Safdar. 2011. "'Father of No One's Son': Abu Ghraib and Torture in the Art of Ayad Alkadhi." *Third Text* 25 (3): 325–34.
Ali, Tariq. 2010. *The Obama Syndrome: Surrender at Home, War Abroad*. London: Verso.

Anoon, Sinan. 2007. *I'jaam: An Iraqi Rhapsody*. Trans. Rebecca C. Johnson and Sinan Anoon. San Francisco: City Lights.

———. 2013. *The Corpse Washer*. Trans. S. Anoon. Cambridge: Yale University Press.

Armitage, Simon. 2008. *The Not Dead*. Hebden Bridge: Pomona.

Barker, Martin. 2011. *A "Toxic Genre": The Iraq War Film*. London: Pluto.

Bilal, Wafaa and Lydersen, Kari. 2008. *Shoot an Iraqi: Art, Life and Resistance under the Gun*. San Francisco: City Lights.

Brothers, Caroline. 1997. *War and Photography: A Cultural History*. London: Routledge.

Burden, Matthew Currier. 2006. *The Blog of War: Front-Line Dispatches from Soldiers in Iraq and Afghanistan*. New York: Simon and Schuster.

Carter, Stephen L. 2011. *The Violence of Peace: America's Wars in the Age of Obama*. New York: Beast Books.

Caruth, Cathy. 1996. *Unclaimed Experience: Trauma, Narrative and History*. Baltimore: Johns Hopkins University Press.

Cavarero, Adriana. 2009. *Horrorism: Naming Contemporary Violence*. Translated by William McCuaig. New York: Columbia University Press.

Chambers, Ross. 2004. *Untimely Interventions: AIDS Writing, Testimonial, and the Rhetoric of Haunting*. Ann Arbor: University of Michigan Press.

Chandrasekeran, Rajiv. 2007. *Imperial Life in the Emerald City*. London: Bloomsbury.

Christini, Tony. 2006. *Homefront*. n. p.: Mainstay Press.

Connor, Steven. 1999. "The Impossibility of the Present: or, from the Contemporary to the Contemporal." In *Literature and the Contemporary: Fictions and Theories of the Present*. Edited by R. Luckhurst and P. Marks, 15–35. Harlow: Pearson.

Costello, Diarmuid and Willsdon, Dominic eds. 2008. *The Life and Death of Images*. London: Tate.

Crosthwaite, Paul. 2005. "'A Secret Code of Pain and Memory': War Trauma and Narrative Organisation in the Fiction of J.G. Ballard". http://www.jgballard.ca/criticism/jgb_secretcode.html (accessed 2 Jan 2014)

Davenport, Lisa. 2009. *Jazz Diplomacy: Promoting America in the Cold War Era*. Jackson: Mississippi University Press.

Dyer, Geoff. 2010. "The Human Heart of the Matter." *Guardian*, Books Review 12 June: 2–4.

Eisenman, Stephen. 2007. *The Abu Ghraib Effect*. London: Reaktion.

Filkins, Dexter. 2009. *The Forever War: Despatches from the War on Terror*. London: Vintage.

Finkel, David. 2010. *The Good Soldiers*. London: Atlantic.

Fountain, Ben. 2013. *Billy Lynn's Long Halftime Walk*. Edinburgh: Canongate.

Franklin, H. Bruce. 1988. *War Stars: The American Superweapon in the American Imagination*. Oxford: Oxford University Press.

Goonan, Kathleen Ann. 2007. *In War Times*. New York: Tor Books.

———. 2011. *This Shared Dream*. New York: Tor Books.

Hamid, Mohsin. 2007. *The Reluctant Fundamentalist*. London: Penguin.

Hanoosh, Yasmeen. 2013. "Beyond the Trauma of War: Iraqi Literature Today". Accessed 2 Jan 2014. http://wordswithoutborders.org/article/beyond-the-trauma-of-war-iraqi-literature-today.

Harris, Paul. 2013. "Emerging Wave of Iraq Fiction Examines America's Role in 'bullshit war'". *Guardian*, 3 January. Accessed 2 Jan 2014. http://www.the-guardian.com/world/2013/jan/03/iraq-fiction-us-military-war

House, Richard. 2013. *The Kills*. London: Picador.

Jafaar, Ali. 2008. "Casualties of War." *Sight & Sound* (Feb): 16–22.

Johnson, Denis. 2008. *Tree of Smoke*. London: Pan.

Junger, Sebastian. 2010. *War*. London: Fourth Estate.

Khadra, Yasmina. 2008. *The Sirens of Baghdad*. Translated by J. Cullen. London: Vintage.

Lim, Bliss Cua. 2009. *Translating Time: Cinema, the Fantastic, and Temporal Critique*. Durham: Duke University Press.

Luckhurst, Roger. 2008. *The Trauma Question*. London: Routledge.

———. 2010. "Beyond Trauma: Torturous Times", *European Journal of English Studies* 14 (1): 11–21.

Marx, Karl. 1981. "The Eighteenth Brumaire of Louis Bonaparte." In *Surveys from Exile*. Edited by David Fernbach, 143–249. Harmondsworth: Penguin.

McEwan, Ian. 2006. *Saturday*. London: Vintage.

Meek, James. 2008. *We Are Now Beginning Our Descent*. Edinburgh: Canongate.

Merino, E. E. and Song, H. R. 2005. "Tracing the Past: An Introduction." *Traces of Contamination: Unearthing the Francoist Legacy in Contemporary Spanish Discourse*. Edited by E. Merino and H. R. Song, 11–26. Lewisburg: Bucknell University Press.

Mustafa, Shakir, ed. 2009. *Contemporary Iraqi Fiction: An Anthology*. Cairo: American University in Cairo.

Myers, B. R. 2007. "A Bright Shining Lie." *Atlantic Monthly* (December). Accessed 2 Jan 2014. http://www.theatlantic.com/magazine/archive/2007/12/a-bright-shining-lie/6434/.

Pascoe, David. 2009. "The Cold War and the 'War on Terror.'" In *The Cambridge Companion to War Writing*. Edited by K. McLoughlin, 239–49. Cambridge: Cambridge University Press.

Peebles, Stacey. 2011. *Welcome to the Suck: Narrating the American Soldier's Experience in Iraq*. Ithaca: Cornell University Press.

Percy, Benjamin. 2007. *Refresh, Refresh*. Saint Paul, Minnesota: Graywolf Press.

Powers, Kevin. 2012. *Yellow Birds*. London: Sceptre.

Pickering, A. 1995. "Cyborg History and the World War II Regime." *Perspectives on Science* 3 (1): 1–48.

Rothberg, Michael. 2009. *Multidirectional Memory: Remembering the Holocaust in the Age of Decolonization*. Stanford University Press.

Shields, David. 2010. *Reality Hunger: A Manifesto*. London: Hamish Hamilton.

Smith, Paul Julian. 2007. "Pan's Labyrinth." *Film Quarterly* 60 (4): 4–9.

Sontag, Susan. 2004. "What Have We Done?" *Guardian* G2 (26 May): 2–5.

Stallabrass, Julian. 2008–9. "Rearranging Corpses, Curatorially." *Photoworks* (Autumn-Winter): 4–9.

Steele, Jonathan. 2011. "The Iraq War's Final End Also Seals Defeat for the Neo-cons." *The Guardian*, 24 October: 28.

Turner, Brian. 2005. *Here, Bullet*. Farmington, Maine: Alice James Books.

Versluys, Kristiaan. 2009. *Out of the Blue: September 11 and the Novel*. New York: Columbia University Press.

Walters, Minette. 2007. *The Chameleon's Shadow*. London: Pan.

Part II
Trauma and the Power of Narrative

4 The Turn to the Self and History in Eva Figes' Autobiographical Works
The Healing of Old Wounds?*

Silvia Pellicer-Ortín

According to the critic Suzanne Keen, it is undeniable that contemporary narratives in English have experienced a historical turn (2006, 167). She alludes in this way to the worldwide literary phenomenon that emerged in the 1970s and 1980s and which Linda Hutcheon defined as "British historiographic metafiction" (1988).[1] Although in Britain this phenomenon emerged in the late 1970s, some critics like Keen have argued that this "historical turn" (167) is still at work in contemporary narratives in English. At the same time, in the last few decades we have observed a proliferation of life-writing in its diverse manifestations together with an increasing academic interest in autobiographical genres (France and St. Clair 2002; Marcus 1994). As I said elsewhere (Pellicer-Ortín 2011, 70), many critics like Roger Luckhurst (2008), Leigh Gilmore (2001), Victoria Stewart (2003), Alison Light (2004) or Sue Vice (2006) have identified a kind of "memoir boom" in the literary scene from the 1990s onwards (Luckhurst 2008, 117). These critics have associated this boom with the contemporary need, both at the individual and the collective level, to narrativise many of the traumatic events that have occurred in the twentieth century, which may explain the large amount of autobiographical narrations published in recent years (132), whose main aim is to expose different hidden traumas.

In the case of British-Jewish literature, both tendencies have developed *pari passu*; especially in writings dealing with the experience of Jewish migrants, targeted at reconstructing Jewish history and identities. In fact, in the 1970s a large group of British-Jewish writers attempted to narrativise their traumatic experiences and their identity conflicts during and after the Holocaust in their fictional and autobiographical creations (Cheyette 1998, xliii-xliii; Brauner 2001, 35; Stähler 2007, 3). As Bryan Cheyette explains, "by the 1970s, the critique of English liberal values, and the rejection of national and communal forms of identification resulted in the location of much British-Jewish writing in a diasporic realm" (2004, 708). Furthermore, as David Brauner points out, Jewish writers have recently begun to be "concerned about the extent of their ignorance about themselves" which has resulted in the considerable number of recent publications raising questions about Jewish history and identity (2001, 34). It is in this context that the autobiographical works of the German-Jewish born British writer Eva Figes powerfully emerge. In his work *Contemporary Jewish Writing in Britain and Ireland: An Anthology* (1998), Cheyette includes Figes within the group of Jewish writers known as "émigrés," together with Ronit Lentin (*Night Train to Mother*, 1989), Dan Jacobson (*The God-Fearer*, 1992), Gabriel Josipovici (*In a Hotel Garden*, 1993) and George Steiner (*The Portage*, 1999). The main feature the works of these writers have in common is their attempt to express their feelings of dislocation and struggle for personhood in their adoptive country, Great Britain (Cheyette 1998, xliii-liii). Further, Figes' autobiographical works should be considered ground-breaking since they anticipated the generation of British-Jewish female writers that started publishing autobiographical and semi autobiographical works dealing with the representation of the Holocaust and their and their families' experiences of migration in the 1990s. Writers such as Linda Grant (*Remind Me Who I Am, Again*, 1998), Anne Karpf (*The War After*, 1996) or Jenny Diski (*Skating to Antarctica*, 1997) followed in Figes' footsteps and attempted to disclose the unveiled feelings of alienation and strangeness they had gone through as immigrant Jews in Britain together with their different traumatic experiences in their autobiographical creations.

Nevertheless, it should be borne in mind that Figes' previous works, published between the late 1960s and 1970s, were very diverse. Her early novels are inheritors of the Modernist stream-of-consciousness interest in the individual and the human mind. This is the reason why Figes was initially regarded as a "representative of contemporary modernism" (Kenyon 1989, 142) and a precursor of experimental fiction in Britain from the 1960s onwards (Byatt 1975, 12). In her early novels, Figes depicted deeply traumatised characters fighting for self-identity. In fictional works such as *Winter Journey* (1967), *Konek Landing* (1969), *Waking* (1981) and *Ghosts* (1988), Figes used a fragmented narrative style in order to portray the disturbed mental processes of these characters and she had recourse to narrative techniques such as fragmentation and dissociation, which many trauma critics, such as Laurie Vickroy (2002), Anne Whitehead (2004), and Ronald Granofsky (1995), have identified as characteristic of trauma narratives. However, in 1978 Figes published her first autobiographical work, *Little Eden: A Child at War*, where she verbalised her and her family's traumatic experiences in the World War II for the first time and where she adopted a more realistic style, abandoning the excessively fragmented aesthetic of her initial work.

The main aim of this article will be to prove that Figes' evolution from her original experimentalism towards a more realistic position, characterized by the attempt at self-definition against the background of history, can fruitfully be addressed from the perspective of Trauma Studies. I will analyse the specific narrative strategies Figes employed in her autobiographical works *Little Eden: A Child at War* (1978a) and *Journey to Nowhere: One Woman Looks for the Promised Land* (2008), in order to represent and understand the traumatic events of her childhood as a survivor of the Holocaust. After analysing these elements, I will address the question of whether the narrativisation of these traumatic experiences, by adhering both to the "turn to history" and the memoir, can be the appropriate means for "heal[ing] old wounds" (Figes 2008, 84), or whether the possibilities of healing are never fully achieved in the works of this writer.

THE BELATED RETURN OF THE PAST

Both *Little Eden* and *Journey to Nowhere* begin with the protagonist's sudden recall of very traumatic events in her early years, activated by the recognition of a long-forgotten place. In the case of *Little Eden*, the revision of the past is motivated by the author-narrator's casual return to Cirencester (the English town Figes' family moved to in 1940) in order to participate in a meeting for young writers. As the author-narrator explains in the opening section:

> I DO NOT know why I never went back before. I saw it marked up on a signpost at odd times over the years, unexpectedly, and each time I was tempted to change my plans for the day and go there instead. [. . .] I craned my head to catch a glimpse of my childhood, buried by years but not forgotten, lost, and now suddenly found. That was ten years ago, but since then the image of myself speeding along that stretch of unchanged road has recurred like a haunting snatch of melody. (7)

In *Journey to Nowhere*, the protagonist's return is to a particular area of London, Lisson Grove. To be more exact, the author-narrator passes by the Samaritan Free Hospital for Women and Children and she starts recalling Edith, the maid Figes' family had before the war and whom she paid a last visit to when Edith was ill in this hospital many years before. This casual encounter brings back the memories of a deeply repressed, unsolved conflict; something which still requires re-evaluation, especially connected with the figure of Edith, and which makes her admit that there is something absent in her current life that should be re-angled: "But the angle of vision changes with time, and at that moment, driving down Lisson Grove half a century later, Edith's story suddenly seemed worth telling. Just because it went against the grain, the in-built prejudices of a lifetime" (2008, 3).

In both cases, the adult author-narrator remembers a period of her childhood and adolescence which had been laid to rest in her unconscious for many years but which has resurfaced to her conscious memory as an adult and needs to be understood for her to continue with her life. In this way, Figes' narrations perform one of the main principles of Trauma Studies. Following the inaugural theories developed by Freud and Breuer in which they explained that a "period of latency" has to elapse between the disastrous accident and the first appearance of the traumatic symptoms in the subject (Freud 2001b, 67–68), Cathy Caruth defined the traumatic experience in terms of its belatedness as a pathology consisting "solely in the *structure of its experience* or reception: the [traumatic] event is not assimilated or experienced fully at the time, but only belatedly, in its repeated *possession* of the one who experiences it. To be traumatised is precisely to be possessed by an image or event" (1995, 4–5). More recently, Roger Luckhurst has emphasised that "this two-stage theory of trauma, the first forgotten impact making a belated return after a hiatus, has been central to cultural trauma theory" (2008, 8). With these ideas in mind, it becomes evident that the main impulse behind these two works is the author-narrator's need to cope with some traumatic past experience which was not fully assimilated or "abreacted" (Freud and Breuer 1991, 59) at the time of the initial shock and which has continued to disturb her life ever since.

These two autobiographical works, then, may be said to respond to the need to verbalise the initial shock as a way to come to terms with it. Belief in the power of voicing of the traumatic experience as a first step in the cure of trauma by transforming the traumatic memories into a coherent narrative was already promoted by such pioneering figures in the field of psychology as Freud and Breuer in their famous theory of the "talking cure" (1991, 57, 68), Pierre Janet (Janet 1928, 438–50) and Carl Jung (1990, 117). Drawing on this, Suzette A. Henke has more recently coined the term "scriptotherapy" to designate the type of life writing that responds to the need "of writing out and writing through traumatic experience in the mode of therapeutic re-enactment" (1998, xii–xiii). As she notes, one of the clearest examples of scriptotherapy takes the form of autobiography since this genre can "so effectively mimic the scene of psychoanalysis that life-writing might provide a therapeutic alternative for victims of severe anxiety and, more seriously, of post-traumatic stress disorder" (xii–xiii). Life-writing becomes, then, a useful tool for the reconstruction of the self after a traumatic process; indeed, the fact that contemporary critics have paid so much attention to these narrative recoveries means that "scriptotherapy has infiltrated the imagination of therapists, literary critics, mental health workers, and narratologists alike" (xiii).

The key traumatic experience that Figes tries to "write through" in these two narrations is the moment in her childhood when she had to leave behind her grandparents and move to Great Britain. This early blow is the origin of such "acting-out" symptoms (Lacapra 2001, 21–22) [2] as recurrent nightmares about the moment of departure from Germany, which become narrative motifs for Figes' representation of trauma in both texts, as the narrators themselves explain:

> But I began to have a recurring dream, in which I regularly relived the moment of departure, the point of no return. The dream was faithful to reality: the plane waiting for take off on the tarmac of Berlin

airport. And a row of abandoned loved ones standing outside the airport building, waving wistfully at survivors whom they could no longer see. (Figes 1978a, 131)

I dreamt sometimes of it, the old world. Always the same dream: the day of departure, a grey March morning, small figures waving from the edge of the airfield as we waited for the plane to take off. (Figes 2008, 10)

However, this is not the only traumatic experience depicted in these works, there are also numerous allusions to the feelings of displacement, loss and guilt experienced by Figes in the course of her life due to her family's forced migration to another country, the terrible consequences the Holocaust had for her relatives and other sectors of society and her difficulties in feeling at ease both with England, the adoptive country, and Germany, her original one. The constant intrusion of the past into present narratives creates two narrations full of digressions and flashbacks characterized by the loss of the "conventional linear sequence" (Whitehead 2004, 6) which, according to critics such as Luckhurst (2008, 88), Laurie Vickroy (2002, 29), and Ronald Granofsky (1995, 17), is one of the main characteristics of so called trauma narratives.

Little Eden: The Past at War

Little Eden (1978a) corresponds to the moment when Figes felt the need to talk about her feelings of alienation as a German-Jewish refugee in Britain for the first time. This work does not follow the pattern of traditional memoir; rather, it interweaves the local history of Cirencester with the ever-present experiences of war and childhood, focalised from the perspective of the adult Figes. As her constant questioning of what it means to be a Jew (74) suggests, the narration may be understood as the author-narrator's quest for identity through the integration of her traumatic experiences as a refugee in a foreign land and the implications of being a Jew during and after the World War II. In her account of her new life in England, the author-narrator makes clear that, despite their efforts to integrate in the new society, migrants, and especially Jews, were consistently treated as different. As she explains: "I had been trying very hard to get myself accepted in the childhood network of streets, school and playground whose laws were strange to me. I was foreign" (13). Throughout the book, the question of foreignness is constantly emphasised (17, 53, 54) despite the fact that, as a child, the author-narrator does not understand what it means to be foreign or a Jew (1, 23, 28, 95). Further, a trace of trauma may be detected in the way she constantly remembers how she was "branded" as different the first day she attended school in England, as can be grasped when she says: "the fact that I had arrived as a foreign child was never forgotten or forgiven, and with the rise of anti-German feeling after the outbreak of war my nationality was always good for abuse" (17).

Throughout the story, the narration of Figes' memories as a child is combined with long passages about the history of England, from the Norman invasions to the present. The author-narrator gives precise historical information about the wars, the different social structures, the succession of monarchs, the folklore and the intermittent phases of the construction of Cirencester cathedral. The historical account is usually evoked when the author-narrator visits a concrete place in the city. For instance, when she goes back to Cirencester Abbey the memory of her mother walking over there also brings to her mind some of the historical events that had happened there during the reign of Henry VIII (1978a, 32), combining in this way both personal and historical memories. The author-narrator believes that each place may become a historical site containing the roots of the population, as she argues in many comments like this one:

> The green fields are rich enough in buried history to keep archaeologists busy for years to come. But life goes on. The old warehouses and stables in Lewis Lane have been pulled down, like everything else, by his lordship, still pumps away at the far end of Lewis Lane, near the old-fashioned flour mill which is still operating [. . .]. In a town that continues to live, the living will continue to make swine-troughs from the remnants of yesterday. But like the people of yesterday, though they have bricked in the arch of the wall behind the lime tree, they have left it standing. If you look carefully enough, as I did, you can still trace out hints of the past. (86)

As these beautiful words suggest, although life goes on, places can talk about the past and unveil many historical episodes, and history will continue to be written generation after generation. People living in the present moment will continue to look at the past, as she is doing now, feeling that they are part of history and giving continuity to the life of this particular town. These words foster the belief that it is not desirable to be stuck in the past but it is necessary to know the past, to see the traces left behind so as to continue writing and understanding our history. In fact, Figes was probably putting into practice these recommendations when she wrote *Little Eden*, thereby enriching her own personal life.

Following this line of thought, it can be argued that, by going back bodily and mentally to the place where some of her traumatic memories were created, the author-narrator rediscovers and assimilates some past events which had been repressed in her unconscious until that moment, as this passage illustrates: "When I first examined my memories of Cirencester there was one image which my adult reason refused to credit even though the vivid picture was indelibly recorded on the retina of my mind's eye" (1978a, 56). In this quotation, the events that have been recorded in

Figes' visual memory refer to some controversial political measures adopted by the Government during the Blitz, like maintaining the traditional fox-hunting season while the small town was being ruined by the continuous German bombings. By returning to this concrete incident of the past, the author-narrator points to the British Government's strategy of pretending that nothing was going wrong during the war. Passages like this one demonstrate that the visual picture of some events of her childhood suddenly comes up on her mental screen. This process allows her now, as a mature person, to understand certain crucial moments of the World War II in England, which were incomprehensible to her when she was a child.

Following Freud's definition of belatedness, it is her return to Cirencester that triggers off simultaneously the need to know and the need to deny some of her traumatic childhood experiences. This rediscovery of memories and history echoes the Freudian observation that the symptoms of trauma start to appear belatedly, after a period of "latency" (Freud 2001b, 67–68). As she puts it, "it was only later, after the brief interlude of time and place was over, that the strains of war and its effect on my family began to impinge on me" (Figes 1978a, 91). Her interest in revising English history has to do with her attempt to obtain the sense of continuity she needs in order to confer a complete meaning on her existence. As she argues, in order to acquire a stable sense of identity as a Jewish immigrant in Britain, she needs to feel the stability and attachment to the English town that can be granted by the history of Cirencester: "[T]he disappearance of Brunel's branch line has done more than deprive me of a dream return to the past as I would have wished it to happen, has done more than disrupt a sense of continuity which otherwise belongs to this small town as the parish church has always belonged to the Market Place" (123). However, she explains that the rediscovery of the local history of Cirencester does not always guarantee this sense of continuity but, as is illustrated in the previous quotation, it makes her more aware of the changes that have taken place in the small town since she was a child. These changes become an echo of the sudden transformations that she has experienced in the course of her life, which have made it more difficult for her to form a complete and stable identity.

At the end of *Little Eden*, after some episodes from the history of England, mainly of Cirencester, have been re-evaluated, and once she has faced the memories of her childhood in the foreign country, the other key traumatic event of her young life is depicted, that is to say, the day when she finally "knew what it meant to be a Jew" (Figes 1978a, 131). The vital learning of the complex meanings and implications of Jewishness took place when her mother told her: "go and see for yourself" (131) and sent her to the Hendon Odeon cinema to watch Belsen's newsreel about the Holocaust, without giving the child any previous explanation. The horror of the images of the war and the concentration camps marked her forever and marked the end of her childhood innocence, which placed her at the beginning of her quest for identity, now from the conscious position of a

Holocaust survivor: "by now I knew myself to be a survivor" (139). There-fore, although the induction into the history of her adoptive country has been essential for the initiation in her process of coming to terms with her past, the traumatic discovery of the Holocaust and the real implications of being a Jew put a dramatic end to her childhood innocence (140). That is to say, the confrontation both with the traumatic Jewish history of persecu-tion and death and the traditional English history, together with the recog-nition of her hybrid Jewish identity that occurs gradually throughout the narration, only meant the beginning of her search for personhood and the first step in the process of healing her trauma, a process that she continued twenty years later in *Journey to Nowhere*.

Journey to Nowhere: One Woman Looks for Reconciliation

In *Journey to Nowhere,* Figes moves on from the combination of history and memoir of *Little Eden* to the blending of autobiography, biography, memoir, history, political essay, history and testimony, in what may be described as a more complex attempt to work through her child-of-war trauma. Together with the personal memories of her life after leaving Nazi Germany, she also recounts the story of Edith, her family maid. After the World War II, Edith decided to migrate to Israel in 1947 hoping to find a land where she could begin a new life, but eventually returned to work for Figes' family in England when she realised that her expectations did not match the reality of the newly created State. Her story illustrates that many Jews who, after having survived the Holocaust in Germany, moved to Palestine hoping to find a kind of Promised Land (Figes 2008, 107) but finding, in fact, a place where they were also looked down on because of their German roots.

The telling of Edith's story shows Figes' need to give voice to some silenced versions of the history of the Holocaust and its after-effects, thus widening the collective dimension of her representation of trauma and moving from the individual representation of her traumatic childhood experiences in England in *Little Eden* to the more collective experiences of alienation undergone by other Jewish immigrant populations in other places in the world such as Germany and Israel. As she has explained, the book's main target is to give a name to all those "countless faces without names who had been part of a vanished world" (2008, 100–101). Further, the testimonial nature of *Journey to Nowhere* is much more complex than the individual testimonial account that readers found in *Little Eden*. This posterior work is composed of various testimonial dimensions which prog-ress from the author's individual story to the collective articulation of other different Jewish experiences, which are mostly represented by the reporting of Edith's past declarations.

In the course of the narration, apart from the genre of testimony, which is not only represented by the testimonial relationship established between

Edith and Eva in 1948 but also by a variety of testimonies of some important socio-political figures of the moment (2008, 125, 127, 162), Figes fuses a variety of genres. Readers can find strictly autobiographical passages that relate the author-narrator's memories, especially in the first section of the book; biographical parts comprehending the main events of Edith's life, mainly in the second section; historical sections that provide important data about the post-Holocaust period, for instance the rise of the Zionist movement (25–29) and the UN politics on the creation of Israel and its consequences (168–75) are portrayed with a wealth of historical detail; and some parts that remind readers of the language of political discourse and essays, mainly in her harsh critique of the US position in the political panorama of the late 1940s, as for example can be seen when the author-narrator offers these categorical statements: "It is difficult to think of any other political decision taken in the twentieth century that has had such long-term and catastrophic consequences, and all for short-term political ends" (176). Passages like this one echo Figes' declarations that she decided to write this book at that precise moment because she felt very angry about Israeli politics in Palestine and so, once she had overcome her initial fears, she decided that this was the time to condemn the political decisions made by the European and American institutions after the war, whose negative consequences are still only too evident (in Pellicer-Ortín 2009, 15). In this sense, it may be stated that *Journey to Nowhere* has a political dimension that did not exist in *Little Eden*. Although both works are representative of Figes' general aim of deconstructing assumed official versions of history in her works by opposing them to the minority perspectives on historical events, *Journey to Nowhere* represents a step further in her attempts to represent the way individual traumatic stories can help shape our conceived versions of history. This is suggested by the author herself when she consciously emphasises that "this is not just a personal story, a memoir of private events; it involves what is now history" (Figes 2008, 139). In keeping with this, Carolina Sánchez-Palencia and Manuel Almagro have identified Figes' use of "petite histoires" in order to deconstruct the so-called "grand narratives" of official history (2000, 113) as a recurrent tendency in her works.

At the same time, in *Journey to Nowhere*, Figes also makes clearer references to the traumatic component of the experiences she and her relatives underwent during and after the World War II. Moreover, she often uses the language of trauma, as when she describes Edith's life experiences as "a *wound* that would not heal" or says that "most of us [the survivors of the Holocaust] are inevitably marked by the *trauma of loss*, though the arrival of a new generation—the grandchildren—is unexpectedly *healing*" (2008, 42; emphasis added). As the author-narrator is now older and has acquired a deeper knowledge of her own self and of the traumatic nature of the events that shaped her life and the society of the moment, the direct allusions to trauma increase as the narration advances. She can now identify the feelings of "guilt" and "fear" (140) that she has suffered all her life and she can recognise the "trauma of loss" (140) and the reason for the compulsive repetition of images that have always persecuted her (58). It is this identification and recognition that allows her to move on to the collective dimension of her own suffering.

Edith's trauma comes to the fore through the author-narrator's recollection of the conversations she had with her in 1948 when Eva was a teenager. For example, the adult narrator puts into words the impressions of her younger self after her conversations with Edith, how Edith had lost all hope and faith after emerging from the Holocaust:

> Looking at the lists of the dead who would never return from Auschwitz, Sachsenhausen or wherever only emphasized what she already knew in her bones: she was quite alone in the world. Nobody loved her, nobody wanted her, no one was looking through those lists in the hope of finding her named as a survivor. (2008, 119)

This passage echoes Edith's realisation once the war was over that she had lost all her relatives and friends during the Nazi regime and she was now alone in the world. Although she was a survivor, she even regretted having lived to tell her tale of the Holocaust; it was at this precise moment that she decided to migrate to Israel. Used to reading the signs of trauma in others, apart from Edith's traumatic testimony, Figes also records evidence of the trauma in other members of her family, like her father's nightmares about the period he was imprisoned in the concentration camp of Dachau (10), her mother's bad and sometimes violent temper towards her daughter (45–51), as well as their reluctance to talk about the painful past—"we did not speak about the dead" (7). In contrast to her family's tendency to obliterate the past without attempting to work through their traumas, the author-narrator makes clear that she needs to "confront [the] ghosts from the past" that have always haunted her (139) in order to get rid of them.

Writing Through the Old Wounds?

Taking all these things into account, some relevant conclusions may be extracted from the analysis of *Little Eden* and *Journey to Nowhere*. In the first place, Figes' autobiographical works respond to the trend of contemporary British-Jewish writings initiated by the "historical turn" that goes back to the decades of the 1970s and 1980s in England, characterized by an attempt to re-evaluate and assimilate a traumatic historical past associated with the Holocaust. Figes' works deserve pride of place within the generation of Jewish writers, especially women, who have recently initiated a "physical and metaphysical journey" to the past (Brauner 2001, 34), aimed at making sense of their selves in history. Moreover, these two works also make evident the memoir boom that has prevailed in the literary panorama of the last decades and which has been closely related to the problematic aspects of the narrativisation of what Figes has described as the "strange" events and times of her childhood (Figes 2008, 1). Eva Figes seems to endorse, then, the vast list of contemporary writers who have felt the need to narrativise their traumatic experiences of migration, survival or abuse in their autobiographical writings (Pellicer-Ortín 2011, 82).

In the second place, *Little Eden* and *Journey to Nowhere* clearly illustrate the key principle of Trauma Studies, which is the power of literature to represent traumatic experiences by having recourse to some specific narrative strategies. The main narrative techniques mentioned by trauma critics and which have appeared in this course of the analysis of Figes' works are digressions, flashbacks, repetitions, dissociation, the recurrent use of images, (Vickroy 2002, 24–32; Whitehead 2004, 86), "the challenge to conventional narrative frameworks and temporality" (Whitehead 2004, 81), and the "multiplicity of testimonial voices" (Whitehead 2004, 88; Gilmore 2001, 7) among others. It may be said that Eva Figes has made use of this variety of techniques in order to represent the paradoxical and aporetic nature of the phenomenon of trauma itself since, in Luckhurst's words, "trauma, in effect, issues a challenge to the capacities of narrative knowledge. In its shock impact trauma is anti-narrative, but it also generates the manic production of retrospective narratives that seek to explicate trauma" (2008, 89). This is the main conflict that predominates in these two works: the contradictory desire of the author-narrator to both encounter and suppress the traumatic events of her childhood. And, as has been demonstrated, it is the need to "explicate trauma" and know the past that wins the inner battle in Figes' case.

Finally, in these two works the author-narrator represents the process of recovering lost or painful memories deeply buried in her unconscious as the first step in the process of writing-healing her trauma. The question that remains to be answered is whether she succeeds in this endeavour. As we have seen, the re-evaluation and assimilation of the history of her adoptive town in *Little Eden* allowed the adult Eva to be conscious of her hybrid position as an expatriate Jew and an English citizen. At the end of this narration, it could be said that her journey towards personhood started at the moment when she accepted her foreign status in England and defined herself as a Jew by acknowledging the history of the Holocaust: "At last I knew what it meant to be a Jew" (Figes 1978a, 131). The writer has declared elsewhere how this knowledge shaped her Jewish identity: "I am Jewish because of what Hitler did; he made more Jews than those that he killed. [. . .] It shaped my whole life and my emotions more than I can ever explain, it goes very, very deep inside" (quoted in Pellicer-Ortín 2009, 10). In *Journey to Nowhere*, Figes manages to represent her life as a refugee in a foreign country through the fusion of literary genres and the depiction of the individual traumatic experience of Edith to illuminate the collective experiences of suffering of Holocaust Jews, in what she has described as an attempt to "make peace with the country in which she was born" (2008, 82).

Little Eden left the possibility of self-healing open for the young Figes once she had learnt about both English and Jewish history, had faced the memories of the events that had been haunting her throughout her life and had admitted that she needed some "time to work through her miseries, loneliness, even homesickness" (1978a, 92), that is to say, once she was aware of the fact that she needed a period of latency to start acknowledging her and her family's traumas. However, in *Journey to Nowhere* the author-narrator still feels the need to reconcile herself with the history of her mother land and to understand the history of the creation of Israel. There are some moments in the narration when readers could assume that the healing function of this life-writing narrative has been fulfilled, as when she reflects on the therapeutic role of writing and transmitting her traumatic life stories to the next generations:

> It was only when I decided to write as a grandmother, remembering my own grandmother, that I found a way in, and the result was curiously comforting. I felt that not only I, but my grandparents, whom I had mourned for so long, were at last at peace. At long last head and heart had synchronized, could function in harmony, not at odds with each other. (2008, 84)

However, although the previous passage could lead to the conclusion that the author-narrator has finally been able to work through the main calamity she constantly acted out in the narrations, i.e., the loss of part of her family during the Holocaust, she admits that a complete victory over the traumatic experience is never possible: "now I am quite open about my past, and have to reassure a younger generation that, really, they have nothing to feel bad about. [. . .] but even that is not always enough to heal old wounds" (84).

Therefore, the direct connection between Figes' autobiographical writings and the healing of her trauma, the efficacy of the process of "scripto-therapy" theorised by Henke, may be confirmed to a certain extent in these two works. The journey this writer started in 1978, when she acknowledged that sometimes: "something happens to remind me that the past is never done with: the shrapnel touches a nerve and I become angry, or unexpected pain brings tears to my eyes" (Figes 1978b, 29), has allowed her to recover the memories of the awful facts of her past, to come to terms with the original traumatic events of her childhood and to voice previously traumatic silenced versions of history. It may be said that both the return to history and the renewed interest in the textual depiction of the traumatised self in these life-writing projects have been successful techniques enabling the author-narrator to create a healing narrative capable of restoring the fragmented self to a position of psychological agency (Henke 1998, xvi). Although the author-narrator acknowledges that the old war wounds will always remain there, silent but always latent, a new integrated self emerges after the process of release and integration of the traumatic past experiences has taken place. The writing-healing process of liberation the author-narrator has performed throughout these two works has enabled her to integrate the painful personal and historical events she had not assimilated at the original moment in her identity. Thus, she has managed to reconcile herself to the past of her adoptive and her mother lands together with the history of the Holocaust in a quite successful way, without forgetting, however, that the traces of trauma will never be totally erased from this survivor's life.

*The research carried out for the writing of this essay is part of a project financed by the Spanish Ministry of Economy and Competitiveness (MINECO) (code FFI2012–32719). The author is also grateful for the support of the Government of Aragón and the European Social Fund (ESF) (code H05).

NOTES

1. The critic Susana Onega has also described this period as characterised by the "eclosion of British historiographic metafiction" (1995, 101).
2. Drawing on the previous Freudian theories about the compulsive repetition of the traumatic event (Freud 2001a, 36–38), Dominick LaCapra coined the term "acting out" to describe the process through which the subject is compelled to relive the traumatic event in an unconscious way as if it were fully present (2001, 70). This process can be manifested in nightmares, the repetition of past events, anxiety, unknown fears, and even self-mutilation.

WORKS CITED

Brauner, David. 2001. *Post-war Jewish Fiction: Ambivalence, Self-Explanation and Transatlantic Connections*. New York: Palgrave.

Byatt, A. S. 1975. "Novel Predicament." *The Times*, March 6: 12.

Caruth, Cathy, ed. 1995. *Trauma: Explorations in Memory*. Baltimore, MD and London: Johns Hopkins University Press.

Cheyette, Bryan, ed. 1998. *Contemporary Jewish Writing in Britain and Ireland: An Anthology*. London: University of Nebraska Press.

———. 2004. "British Writing and the Turn towards Diaspora." In The Cambridge History of Twentieth-Century British Literature. Edited by Laura Marcus and Peter Nichols, 700–15. Cambridge: Cambridge University Press.

Diski, Jenny. (1997) 2006. *Skating to Antarctica*. London: Virago.

Figes, Eva. 1967. *Winter Journey*. London: Faber and Faber.

Figes, Eva. (1969) 1972. *Konek Landing*. London: Panther Books.

———. 1978a. *Little Eden: A Child at War*. New York: Persea Books.

———. 1978b. "The Long Passage to Little England." *The Observer*, June 6: 29.

———. 1981. *Waking*. London: Hamish Hamilton.

———. 1988. *Ghosts*. London: Hamish Hamilton.

———. 2008. *Journey to Nowhere: One Woman Looks for the Promised Land*. London: Granta.

France, Peter, and William St. Clair, eds. (2002) 2004. *Mapping Lives. The Uses of Biography*. New York: Oxford University Press.

Freud, Sigmund, and Josef Breuer. (1895) 1991. *Studies on Hysteria*. Edited and translated by James and Alix Strachey. London: Penguin.

Freud, Sigmund. (1920) 2001a. "Beyond the Pleasure Principle." In *Beyond the Pleasure Principle, Group Psychology and Other Works. The Standard Edition of the Complete Psychological Works of Sigmund Freud* Vol. XVIII. Edited and translated by James Strachey, 7–64. London: Vintage.

———. (1939) 2001b. "Moses and Monotheism." In *Moses and Monotheism: An Outline of Psycho-Analysis and Other Works. The Standard Edition of the Complete Psychological Works of Sigmund Freud* Vol. XXIII. Edited and translated by James Strachey, 1–137. London: Vintage.

Gilmore, Leigh. 2001. *The Limits of Autobiography: Trauma and Testimony*. Ithaca and London: Cornell University Press.

Granofsky, Ronald. 1995. *The Trauma Novel: Contemporary Symbolic Depictions of Collective Disaster*. New York: Peter Lang.

Grant, Linda. 1998. *Remind Me Who I Am, Again*. London: Granta.

Henke, Suzette A. 1998. *Shattered Subjects: Trauma and Testimony in Women's Life-Writing*. London: McMillan.

Hutcheon, Linda. 1988. *A Poetics of Postmodernism: History, Theory, Fiction*. New York and London: Routledge.

Jacobson, Dan. 1992. *The God-Fearer*. London: Hodder and Stoughton.

Janet, Pierre. 1928. *L' evolution de la mémoire et la notion du temps*. Paris: Cahine.

Josipovici, Gabriel. 1993. *In a Hotel Garden*. Manchester: Carcanet.

Jung, Carl G. (1959) 1990. *The Archetypes and the Collective Unconscious. Completed Works*, Vol. 9, part 1. Edited by M. Fordham and G. Adler. Translated by R. C. Hull and H. Read. London: Routledge.

Karpf, Anne. (1996) 2008. *The War After: Living with the Holocaust*. London: Faber and Faber.

Keen, Suzanne. 2006. "The Historical Turn in British Fiction." In *A Concise Companion to Contemporary British Fiction*. Edited by James F. English, 167–83. Oxford: Blackwell.

Kenyon, Olga. 1989. "Eva Figes." In *Women Writers Talk: Interviews with 10 Women Writers*, 69–90. Oxford: Lennard.

LaCapra, Dominick. 2001. *Writing History, Writing Trauma*. Baltimore and London: The Johns Hopkins University Press.

Lentin, Ronit. 1989. *Night Train to Mother*. Dublin: Attic Press.

Light, Alison. 2004. "Writing Lives." In *The Cambridge History of Twentieth Century Literature*, Vol. 1. Edited by Laura Marcus and Peter Nicholls, 751–67. Cambridge: Cambridge University Press.

Luckhurst, Roger. 2008. *The Trauma Question*. London: Routledge.

Marcus, Laura. 1994. *Auto/biographical Discourses: Theory, Criticism, Practice*. Manchester and New York: Manchester University Press.

Onega, Susana. 1995. "British historiographic metafiction." In *Metafiction*. Edited by Mark Currie, 172–78. London and New York: Longman.

Pellicer-Ortín, Silvia. 2009. "Interview with Eva Figes." Unpublished.

———. 2011. "Testimony and the Representation of Trauma in Eva Figes' *Journey to Nowhere*." *Atlantis: Journal of the Spanish Association of Anglo-American Studies* 33 (1): 69–84.

Sanchez-Palencia, Carolina, and Manuel Almagro. 2000. "'My Father's Daughter': Deborah Milton's Biography of Silence." *Journal of English Studies* II: 113–24.

Stähler, Axel. 2007. *Anglophone Jewish Literature*. Oxon: Routledge.

Steiner, George. 1999. *The Portage*. Chicago: University of Chicago Press.

Stewart, Victoria. 2003. *Women's Autobiography: War and Trauma*. London: Palgrave Macmillan.

Vice, Sue. 2006. "Writing the Self: Memoirs by German Exiles, British-Jewish Women." In*"In the Open": Jewish Women Writers and British Culture*. Edited by Claire M. Tylee, 189–209. Newark: University of Delaware Press.

Vickroy, Laurie. 2002. *Trauma and Survival in Contemporary Fiction*. Virginia: University of Virginia Press.

Whitehead, Anne. 2004. *Trauma Fiction*. Edinburgh: Edinburgh University Press.

5 History, Dreams, and Shards

On Starting Over in Jenny Diski's *Then Again*

Gerd Bayer

The contemporary British writer Jenny Diski has always been willing to take on the difficult, the troublesome, and the formally demanding. Not unfamiliar with taxing mental states herself, she has written a number of books that have strained psyches as one of their central topics. For instance, her first novel, *Nothing Natural* (1986), portrayed a single mother who gets drawn into an abusive sexual relationship; her 2008 novel, *Apology for the Woman Writing*, presents the life of Marie de Gournay, the adopted daughter of French essayist Michel Montaigne, who herself wrote a novel that has as its protagonist a rather obsessive woman. Generational tensions and mental conflicts truly abound in Diski's oeuvre. And adoption frequently plays a central role for the transmission of culture in her works, evoking the tropes of fragmentation and memory. In this chapter, I concentrate on Diski's fourth novel, *Then Again*, and argue that in its approach to trans-generational memory it presents artistic and aesthetic work as a countermeasure to traumatic moments of crisis, offering at the same time a potential approach to the past that addresses traumata without risk of causing secondary pain.

First published in 1990, the novel combines the lives of three protagonists: first, a medieval Esther, who is the only person to survive an anti-Semitic pogrom. She is subsequently adopted by a Christian family but later becomes victimised and sexually abused by a priest who exploits both her intelligence and her unconscious knowledge that she does not fully belong to her community. The priest seems envious of her gift for religious belief, maliciously accuses her of witchcraft, and has her burned at the stake. This medieval Esther exists only in the dream world of the second protagonist, a modern Esther, who feels psychically connected to her earlier namesake. A single mother and pottery designer, this contemporary Esther also struggles with her own sense of identity and, at the opening of the book, has to respond to the disappearance of her daughter, Katya, the third main character in *Then Again*. Katya has suffered a mental breakdown following a lengthy process of religious soul searching. She tries to find out the truth about God and the devil, about right and wrong, and as a consequence finds herself walking through London in a very confused state of mind, ending up being badly treated both by abusive men and by mental institutions. Much of Esther's dialogue in the novel deals with her attempts at finding proper treatment for Katya, with the mother being rather unwilling to have her daughter committed to an institution, somehow realising that the two of them need to work out the underlying issues that caused this crisis.

As this short plot survey already indicates, Diski's novel connects questions of memory, trauma, family, institutions, and art. Through modern Esther's work as a pottery designer, Diski inserts into the fiction a creative means of manipulating material reality, and thus of actively engaging with the past outside conventional norms and formats. In fact, the consequences of an insufficient insertion of the past in the present form the ethical backbone of *Then Again*. Like other recent Holocaust fiction, Diski's novel draws heavily on the postmodern tropes of fragmentation and self-critical writing, following what Ephraim Sicher defines as a distinctive marker of recent fiction, distinguishing it from earlier literary attempts "to represent atrocity by employing conventional means of realism" (2005, 176). This seconds a point also made by Michael Rothberg in his book about traumatic realism and postmodern Holocaust art. Rothberg (2000) and others have argued that the aesthetic principles of postmodernism can be used productively to deal with ethical and traumatic issues (see Eaglestone 2004; Bayer 2009). Diski's work underlines that idea by uniting technical features that belong to postmodernism with a discussion of the Holocaust and trans-generational memory.

Marianne Hirsch has famously introduced the concept of postmemory into Holocaust discourse (1997). Her research into the way that memories, often triggered by photographs, are passed on to later generations has drawn attention to the way that the forces of trauma can affect even those born after the original traumatic events have occurred. Concentrating on the twentieth-century Nazi-Holocaust of European Jewry, Hirsch's work can nevertheless be applied to other historical moments. Its focus on the second generation requires at the same time a critical inquiry into how later generations will be affected by postmemory, and whether memories may affect the third or later generations in a different manner (Bayer 2010).

Then Again also addresses trans-generational memory, but it does so by asking to what extent the pain of suffering induced by postmemory should be entirely subdued. By confronting a mother who has repressed her traumatic trans-generational memories of the Holocaust with the difficult choice of how to respond to her daughter's mental breakdown—therapy and drugs being the two options—Diski to some extent asks whether traumatic events open a pathway for an alternative approach to past and memory that complements the dominant modes of repression or amnesic atonement. In this, she moves close to what Roger Luckhurst critically diagnoses as W. G. Sebald's "traumatophilia," but unlike the Anglo-German author, Diski cannot truly be accused of "taking a kind of perverse delight in the repetition or abject assumption of a collapsed trauma subjectivity" (Luckhurst 2008,

111). In fact, nothing could be further from Diski's mind than a celebratory approach to mental pain and suffering: she has experienced extreme mental states herself and has written in her non-fiction, for instance in *Skating to Antarctica* (1997), about the pain and anxiety she experienced and the need for quality support in such situations.[1]

In *Then Again* Diski nevertheless considers psychological states outside the norm as potentially beneficial for the overall mental balance of any culture. Paying a high price, the tortured protagonists in her novel alleviate the cultural discontents of post-Freudian civilisations. While Diski's first novel, *Nothing Natural*, presents sadism and masochism as potentially harmful but nevertheless normal behaviours, her later novels reverse this argument and instead ask how "normal" culture stands to benefit from the psychic processes that those suffering mental pain experience. She thus works in a mode that closely mirrors the argument presented by Gilles Deleuze and Félix Guattari in their *Anti-Oedipus*, a work that embraces alternative rationalities and denounces the at times violent approaches to be found in psychiatric circles, presented by the two critics as panoptic cultural police. While clearly not celebrating trauma, Diski nevertheless insists on the necessity of engaging with topics such as the Holocaust, regardless of the pain this may cause: the implicit argument being that the silencing of trauma has even worse consequences in the long run. In addressing this issue, Diski follows a pronouncement by Theodor Adorno, who, in his essay on Bertolt Brecht and commitment in art, observes: "When genocide becomes part of the cultural heritage in the themes of committed literature, it becomes easier to continue to play along with the culture which gave birth to murder" (1995, 5). In other words, the kind of traumatic presence conjured up in Diski's novels provides a much-needed counter-narrative to the dominant modes of celebrating positive historical moments and, alternatively, of amnesia and repression. In *Then Again*, Diski reminds readers of past murders, and she does so while simultaneously creating a story of conflicting discourses and all-too-powerful systems of answers.

The book's title, *Then Again*, alludes through the rhetorical evocation of ambiguity to the troubled status of traumatic memory. Diski in this joins critics like Zygmunt Bauman, who has noted with respect to the Holocaust that there is "No logically coherent ethical code" that can be used to represent "the essentially ambivalent condition of morality" (1993, 10). Diski's *Then Again* evokes this aporia in its title and goes on to apply it to the transmission of traumatic memories across generations. Both a burden and a mission, trans-generational trauma yearns to be overcome and, then again, refuses to dissolve into amnesia. The circular back and forth implied by this mnemonic movement correlates to the impossibility of forgetting what, at heart, is unbearable to remember. Diski's novel clearly suggests that the consequences are severe and make their impact felt over generations, with the possibility of reaching a point of almost complete debility. The crises induced by such memories, Diski evokes in Esther's escapist dreaming and Katya's manic religious soul-searching. In a similar novel, Thane Rosenbaum (1999) has the protagonist of *Second Hand Smoke* relate to his parents' trauma through a heightened sense of victimisation and a violent commitment to retribution. By fighting the very Nazism that caused his parents' suffering, the offspring in that novel almost succumbs to the cancerous growth of complete identification with a historical moment massively over-inscribed by extreme forms of evil.

Diski's Esther avoids this fate by transposing her traumatic pain into creative production: she translates insecurity into productivity, albeit leaving aside the ethical implications of her transformations. Her engagement with medieval anti-Semitism and murderous pogroms is restricted to the dream stage, the removed reality of the other Esther that she keeps clearly separate from her waking self. And yet, the novel suggests, the fear that unavoidably arises when the mind touches on horror surfaces in her daughter's struggle with God and her insecurity with respect to her own ethical identity. What Diski's novel envisions is the Freudian return of the repressed as it plays out in the context of trauma, what Cathy Caruth has described as "the literal return of the event against the will of the one it inhabits" (1995, 5). However, the novel does not imply that Katya or even Esther had been directly exposed to trauma: it suggests that there may be different reasons for becoming host to a traumatic dream.

Diski thus adds a new facet to the memorialising strategies of cross-generational memory. While the kind of postmemory discussed by Marianne Hirsch concentrates on the second generation and its difficult relationship to the pain suffered by its parents (McGlothlin 2006), Diski envisions a situation where the silencing of such a memory fatefully disconnects the narrative from the pain. Or, to put this differently, she creates a fictional scenario where a parent who is herself not a direct victim nevertheless inadvertently passes on the legacy of trauma, without, however, offering her daughter a narrative frame in which to insert the emotional distress.[2] Unlike J. M. Coetzee's eponymous alter ego in *Elizabeth Costello* (2003), who fears that too close an encounter with fictionalised representations of evil may irredeemably taint the soul, Diski's tale of an n'th-degree inheritor of pain who is not openly informed about her unspoken mnemonic inheritance has the adverse effect of causing mental distress in her own daughter through the very lack of a narrative framework or of explicit verbalisation.

This kind of silencing resonates with other works by Diski (Terrien 2009), but it furthermore evokes the psychoanalytical research by Nicolas Abraham and Maria Torok, collected in *The Shell and the Kernel* (1994). Abraham in particular theorises how memories can be passed on to later generations despite the fact that nothing was said about a particular incident. He describes this spectral phenomenon as a phantom and notes that "It passes—in a way yet to be determined—from the parent's unconscious into the child's." Periods of calm silence, with the traumatic past buried in the unconscious, are interrupted by the appearance of such buried memories, a process that Abraham describes as a "periodic and compulsive return" (Abraham and Torok 1994, 173). He is quite explicit about the circumstances that bring such a phantom into being, arguing that the kind of mental suffering it causes goes back to a situation when "a gap was transmitted to the subject with the result of barring him or her from the specific introjections he or she would seek at present" (174). It is rather uncanny to note that Abraham's description of the reasons why such phantoms get activated fits almost perfectly Diski's use of painful memories in *Then Again*. In words that seem to describe both Esther, the orphaned mother, and her daughter Katya, Abraham defines the underlying forces at work with this sentence: "The presence of the phantom indicates the effects, on the descendants, of something that had inflicted narcissistic injury or even catastrophe on the parents" (174). With respect to Diski's novel, the injuries and catastrophes experienced by the respective parents do indeed surface in the painful and traumatising visions of their children: Esther's dream about the suffering of the medieval Esther may thus be read as a condensed return of both her biological and adoptive parents' personal life experiences. And Katya's soul-searching repeats her mother's attempt at finding her own place in the generational matrix of ancestry and cultural memory. As neither of them can claim solid knowledge about their parental experiences and pain—silence having prevented that kind of postmemory to develop—their exposure to mental turmoil could in fact be explained as the kind of return of a spectral phantom as described by Abraham, who himself turns to aesthetic works for an explanation of this dynamic. In an essay on *Hamlet*, he suggests that one reason why Shakespeare's play has found such lasting interest is that it sets out "to spur the public to react unconsciously to the enigmas that remain." *Hamlet*, much like Diski's novel (and Holocaust art in general), wants "to cancel the secret buried in the unconscious and to display it *in its initial openness*." One way of silencing the phantom is by "reducing the sin attached to someone else's secret and stating it in acceptable terms" (189). What this boils down to is a need to verbalise and thus make openly visible the lingering pain of past experiences even across generations.

Other aspects of Diski's novel also toy with the concept of postmemory: Esther, Katya's mother, is the adopted daughter of a college-teaching couple named Dinah and Geoffrey Friedman. Even though their religious background is never specified—they self-identify as atheists and lower-case humanists—their name suggests Jewish roots. Much of their energy goes into global charity work, with some coded allusions to survivor's guilt: "One did what one could to alleviate suffering wherever it occurred. But the point was that it occurred everywhere. Who escaped? If the Friedmans were exempt from most of it, their pain lay precisely in their exemption and their awareness of the misery of others" (Diski 1990, 27). Esther, for her part, speculates that her birth may have been connected to the presence of US soldiers in the UK in 1945 (25), but she clearly identifies with her parents' life style and ethical choices. As a logical consequence, she later starts a PhD "on the psychological aspects of a series of Jewish massacres that occurred in the mid-fourteenth century" (25), a project she quickly abandons after her parents' death. Blood ties, however, provide a lasting connection to the social commitment carried out by her adoptive parents, while her lack of knowledge about her biological parents leaves her permanently "*wondering*" about her true identity (24).[3] The dreams connecting her to the medieval Esther provide a condensed image of such identification in that they include her parents' commitment to suffering, her own work on pogroms, and the unspoken religion of her adoptive parents. In the return of this phantom, Esther (and to some extent even Katya) confronts the unspoken pain of her ancestors, both biological and adoptive.

Esther's name furthermore brings up a range of condensed references: the Biblical Esther evokes Jewish pogroms, adopted daughters, sexual exploitation, but also the power of the spoken word to prevent disaster. It is also closely connected to the Jewish celebration of Purim.[4] In fact, the etymology of "Esther" includes "star" and thus forms a direct link to the Latin "aster" (star) in disaster. The searching for identification that marks Diski's Esther thus seems to be connected to questions of generational responsibility but also to larger issues of ethical memory. While the novel does not mention whether Esther's adoptive or biological parents were personally affected by the Nazi Holocaust, the spectre of that genocide nevertheless haunts the book. In effect, it breaks through both in Esther's dream world and, arguably, in her daughter's searching for an ethical centre. The mysterious circumstances surrounding Esther's birth in 1945 provide a further link to the Second World War, allowing for a reading that connects her biography to the sinister developments playing out at that historical moment. What all these various gaps in memory appear to have caused is the passing down to later generations of a traumatic phantom that will occasionally rise to haunt its hosts.

It is a little easier to disentangle the pain suffered by Esther's daughter. Katya's psychological trouble starts when she has what the book repeatedly describes as an experience of "grace" (Diski 1990, 107). She awakes one night to rainfall and stormy winds that constitute "a *force*" that "filled her with—with the indescribable. With fullness. With . . . *grace*" (108). While Katya experiences this moment as a religious epiphany, readers may also see in this a kind of sexual awakening, a reading that Diski seems to toy with when writing that Katya feels "a fluid rush that eventually broke through her skin and joined her to the universe" (108). The narrator, however, soon passes a final verdict—one that closely resembles Katya's own interpretation of the event—and notes laconically: "Katya, at fourteen, had seen God dancing in the particles of the universe, and danced along in God's midst" (108). The confusion this causes her is explained by the fact that Esther had brought up her daughter without orthodox religious beliefs, instead mixing Biblical, mythical, paleo-anthropological and other accounts of creation and the underlying force that has created and keeps alive human and other existence. In short, "Nothing had taught her what to do with *belief* when it had come and pressed on her heart and mind" (111). As a consequence she tries to purify her whole existence, refuses to eat, and gives away what used to be her favourite clothes and toys.

To some extent, Katya can be described as undergoing a process of atonement without, however, being able to account for her misdeed, unless it be that of not accepting belief earlier. An attempt to tell her mother about her experience of a divine *unio mystica* fails because Esther cannot (or will not) take her daughter's religious feelings seriously. This in turn, Diski's novel implies, goes back to Esther's own attitude towards organised religions and the fact that she associates them mainly with control, violence, and the persecution of non-believers. As if to prove this, Katya's attempt to talk to a priest at a local church is a complete disaster, as the young priest reacts angrily when Katya claims to have had personal communication with God, something that the envious cleric will not accept. He instead claims that her arrogance in assuming such privileged treatment is a clear indication of her own wickedness and sinfulness. Katya, mentally unstable as she is at that moment, immediately accepts this explanation and falls into a serious fit of self-loathing, believing that "She was infested with wickedness so palpable that her flesh crawled" (Diski 1990, 125). This constellation—a representative of official religions instilling a sense of guilt—not only sets her on the path towards an almost complete mental breakdown but it also uncannily echoes the experience that the medieval Esther undergoes when she approaches her priest with her religious feelings and needs. What *Then Again* thus implies is that the dynamics of institutional control have remained fundamentally unchanged and that those in need of support and help are frequently exploited and pushed further into dependency. The constellation is of particular interest if seen through the prism of intergenerational memory, as it is Esther, Katya's mother, who has these dreams of the medieval Esther but does not communicate to her own daughter the kind of information that might have helped her to avoid such a drastic mental decline and could have warned her about the evil streak that runs through many institutions.

All these textual features support a reading of the novel that sees it as an implicit denunciation of the dangerous consequences that the silencing of trauma can have, even across generations. Despite remaining ignorant about their ancestral history, both Esther and Katya are the kind of "adoptive or intellectual witnesses" about whom Geoffrey Hartman notes that they are people "of conscience who feel that what happened is not an epochal historical event equivalent to other such events" (2006, 260). Diski's novel— unsurprisingly, given her fascination with psychoanalysis—thus not only substantiates Abraham's writing about the phantom and its cross-generational spectral appearance, but it also more generally evokes the powerful dynamics of an almost Jungian collective unconscious from which cultural memory draws an awareness of traumatic pasts.[5] Reminiscent of the force at the heart of Jung's approach to therapy, her protagonists undergo mental duress as a consequence of failing to give sufficient prominence, in their consciousness, to the memory of anti-Semitic pogroms. The tension finally and fully breaks out in Katya, a third (or even n'th) generation survivor of the twentieth-century Holocaust, and thus even further removed from the medieval pogroms that haunt her mother.

Diski's novel ends, quite appropriately, without offering closure. Katya's future remains quite as unclear as the relationship between her and her mother. However, Esther at the end of the novel has gone through some changes. In a final conversation with Ben, her therapist friend and lover, Esther realises that her daughter's mental instability to a substantial degree grew from Katya's expectation that she would find clear answers to questions about good and evil. Those answers, Esther is now ready to admit (with Zygmunt Bauman), might not be forthcoming, which does not mean that asking the questions is futile. Rather, in a nod to poststructuralist notions of epistemology, Esther and Ben agree that resistance to large explanatory systems is a necessary feature of (post)modern life. In fact, when Esther notes that "The Inquisition stopped people from asking questions," she is reminded by Ben that "the questions go on even if the people don't" (Diski 1990, 209), his comment providing further testimony to the spectral existence of some memories. The novel links the extreme violence that grew from the medieval institution of the Inquisition to a Foucauldian system of omnipresent discourse control. Non-conforming views were punished by death, explaining at least in part the 'rationale' that drove the medieval pogroms evoked in the novel. Such pogroms were directed at disbelievers, in the sense of those who were unwilling to subordinate themselves to the discursive models set by the majority. In the book's frequent allusions to psychiatry's all too swift resorting to drugs as a means of stopping a patient's mind from engaging in inappropriate thinking or questioning, a certain parallelism between the Inquisition and medical psychiatry is drawn. While Diski thus implicitly subscribes to the anti-Oedipal scepticism found in Deleuze and Guattari's two volume *Capitalism and Schizophrenia* (and thus follows a similar strategy to the one that Peter Shaffer applied in his 1973 play *Equus*), her final focus is less on recent developments within the treatment of mental non-conformists, and more on the issue at the heart of the questions that Katya asks.[6] The novel in fact insinuates that asking ethically demanding questions about traumatic pasts is both necessary and risky, in particular since they might not produce easy answers.

Furthermore, the answers may not necessarily be verbal or communicable in standard formats. As an alternative, Diski evokes the power of art to provide an avenue towards understanding. The novel clearly places special emphasis on how creative and aesthetic work relates to Esther's process of both learning and healing: soon after her dialogue with Ben about the inability of the Inquisition to stop questions from being asked, Esther designs yet another plate, this time opting for a different strategy. She takes fully painted plates and drops them on the floor. The resulting shards she reassembles, creating a design that is built on broken pieces. Picking up the pieces, aesthetically, turns out to be hard work, since she has to paint very carefully along the borders of two colours: "With each change of colour she had to wait for the adjacent colour to dry enough not to bleed when she applied the new one" (Diski 1990, 211). Metaphorically commenting on the process of healing—the "bleeding" obliquely evokes a wound and hence, etymologically, a trauma—this scene also insists on documenting the lines of fissures, on keeping the ruptures alive, and on carefully distinguishing between particular aspects. It employs typically postmodern stylistic innovation and fragmentation as a means for dealing with traumatic memories (Reiter 2005, 194). The resulting design, called "Then Again," Esther nevertheless realises no one will buy as it visualises both pain and healing and hence touches on issues most people prefer not to be reminded of. The name she chooses for her new plate design also implies the kind of resistance mentioned earlier with respect to the Inquisition. In the case of the painted plates, "then again" alludes to how even fragmented shards—metaphorically evoking memories of trauma—can be reassembled, given enough time, love, and patience.

This recipe, however, Diski contrasts sharply with the strict submission to traditions that culture habitually demands from members of any community. Having closed the penultimate chapter with the pottery design just discussed, Diski opens the final chapter with a quotation from the *Shema, ysrael,* the Torah lines that have the purpose of commanding daily observance through prayer. Initially interspersing the Hebrew phrases with Esther's thoughts, the chapter later offers a translation. Commenting on the prayer's commandment concerning dutiful observance, Esther notes that "She could not praise the Lord with all her heart and soul and might," the reason being that "the questions were upon her heart" (Diski 1990, 213). Despite realising that "she could not pass on unquestioning faith," she nevertheless accepts the generational responsibility of teaching about the past in that "*something* should be passed on: what is known, what is understood; what is not understood and cannot be, but can be asked" (214). In a sense, Esther blames herself for not addressing earlier the questions that her daughter had ended up asking. In other words, she realises that she did not sufficiently honour the obligation to remember the past. Through her dream relationship with the medieval Esther, she had come to realise that when faced with the immensity of disaster, identification with an individual story can help turn trauma into narrative. In order to pass memory on—and to prevent it from exerting the full force of what Thane Rosenbaum describes in an essay on Holocaust art as the past's gift of a "poisonous fruit" (2006, 491)—it needs to be properly emplotted first. Spectral memories of past phantoms, Diski implies, have to be detoxicated through verbalisation and aesthetic representations, even when there is no clear answer or response to every aspect related to the pain. As Esther begins to face both her personal past and the troubling questions that her daughter asks as a consequence of her own failure to provide ethical guidance, she formulates a mnemonic program that uncannily echoes the strategy picked up later (3 years, to be precise) by the US Holocaust Memorial Museum. Diski suggests: "Try thinking about one person's pain. Try a single agony. Acknowledge it. Add it to the unfinished sum that would, in the end, *at* the end, become a total of what had been understood" (1990, 216). Final answers might not be forthcoming but—then again—asking the questions might help future generations in dealing with the trauma that gave rise to them.

In the light of what this chapter has discussed, it can be concluded that Diski's *Then Again* abounds in unfinished business, unanswerable questions, elusive answers, and desperately required empathy (Koopman 2010). It contributes to the academic discussion about trauma and narrative in that it first of all emphasises how important it is to put pain into narrative in order to make it communicable—and thereby to prevent it from turning into the kind of hurtful phantom described by Abraham—and how impossible it may well be to find a narrative that redeems both its teller and listener of all pain. Similar to some current discussions in what is known as the medical humanities, a field that applies humanities research to the medical profession (Charon 2008; Schnell 2004), Diski's novel abstains from aiming at final answers and instead propagates an openness towards asking the right questions, in particular in situations where they may well remain unanswered. Such asking and telling may frequently violate the conventional borders of polite conversation, the traditions within some institutions, or even the core of some organised religions, but Diski presents it as a necessary step in addressing the kind of questions that will always eventually resurface. Arthur Frank has argued that a person who suffers extraordinarily may well have receded into an inner world structured by chaos, making "it difficult to communicate" (Frank 2004, 219), yet also emphasises that communication remains paramount, with skilful communicators drawing not so much on standardised knowledge but on insightful and often metaphorical approaches gleaned from the indirect teaching of creative works. Just as art may help to reassemble the past's broken pieces, so it may simultaneously prevent traumatic memories from travelling the trans-generational and spectral pathways that, in Diski's *Then Again*, lead to severe mental distress in the descendents of the very people who tried to be protective by staying silent.

NOTES

1. In *Don't* (1998), her first collection of book reviews and short (non-fictional) personal essays, Diski frequently returns to issues of madness and suffering, to the creative spirit, but also to the memories and legacies of the Holocaust.
2. On post-memory, memory, literature, and trauma see Hirsch (2008), but also Assmann (2003) and Bayer (2009).

3. On the difficult role that truthful representations of reality play within the context of trauma, see Felman and Laub (1992, 60–63), who point out that deviations between memory and historical facts do not compromise the act of giving testimony. Robert Eaglestone has meanwhile pointed out that *"memory without identity is meaningless"* (2004, 75).
4. Esther is queen to Ahasverus, a non-Jewish Persian king who prevents a Jewish genocide but whose name later became associated with the myth of the Wandering Jew. On "disaster," see also Maurice Blanchot (1995).
5. For further reading on Jungian psychoanalysis, trauma theory, and Pierre Janet's claim that traumatic experiences need to be narrated in order to heal, see Jean-Michel Ganteau and Susana Onega, introduction to *Ethics and Trauma in Contemporary British Fiction* (2011).
6. See Crawford and Baker (2009), who discuss some crucial aspects of the role that madness plays in Diski's work.

WORKS CITED

Abraham, Nicolas, and Maria Torok. 1994. *The Shell and the Kernel: Renewals of Psychoanalysis*. Edited by Nicholas T. Rand. Chicago: University of Chicago Press.

Adorno, Theodor W. 1995. "Commitment." In *Aesthetics and Politics*. Edited by Fredric Jameson, 177–95. London: Verso.

Assmann, Aleida. 2003. "Three Stabilizers of Memory: Affect–Symbol–Trauma." In *Sites of Memory in American Literature and Culture*. Edited by Udo J. Hebel, 15–30. Heidelberg: Winter.

Bauman, Zygmunt. 1993. *Postmodern Ethics*. Oxford: Blackwell.

Bayer, Gerd. 2009. "Der Holocaust als Metapher in postmodernen und postkolonialen Romanen." In *Literatur und Holocaust*. Edited by Gerd Bayer and Rudolf Freiburg, 267–90. Würzburg: Königshausen & Neumann.

———. 2010. "After Postmemory: Holocaust Cinema and the Third Generation." *Shofar* 28 (4): 116–32.

Blanchot, Maurice. 1995. *The Writing of the Disaster*. Translated by Ann Smock. Lincoln: University of Nebraska Press.

Caruth, Cathy. 1995. "Introduction." In *Trauma: Explorations in Memory*. Edited by Cathy Caruth, 3–12. Baltimore: Johns Hopkins University Press.

Charon, Rita. 2008. *Narrative Medicine: Honoring the Stories of Illness*. Oxford: Oxford University Press.

Coetzee, J.M. 2003. *Elizabeth Costello*. London: Secker and Warburg.

Crawford, Paul, and Charley Baker. 2009. "Literature and Madness: Fiction for Students and Professionals." *Journal of Medical Humanities* 30: 237–51.

Deleuze, Gilles, and Félix Guattari. 1984. *Anti-Oedipus: Capitalism and Schizophrenia*. Translated by Robert Hurley, Mark Seem, and Helen R. Lane. London: Athlone.

Diski, Jenny. (1990) 1998. *Then Again*. London: Granta.

———. 1997. *Skating to Antarctica*. London: Granta.

———. 1998. *Don't*. London: Granta.

———. 2008. *Apology for the Woman Writing*. London: Virago.

Eaglestone, Robert. 2004. *The Holocaust and the Postmodern*. Oxford: Oxford University Press.

Felman, Shoshana, and Dori Laub. 1992. *Testimony: Crises of Witnessing in Literature, Psychoanalysis and History*. New York: Routledge.

Frank, Arthur W. 2004. "Asking the Right Question about Pain: Narrative and *Phronesis*." *Literature and Medicine* 23 (2): 209–25.

Ganteau, Jean-Michel, and Susana Onega. 2011. "Introduction." In *Ethics and Trauma in Contemporary British Fiction*. Edited by Susana Onega and Jean-Michel Ganteau, 7–19. Amsterdam: Rodopi.

Hartman, Geoffrey. 2006. "The Humanities of Testimony: An Introduction." *Poetics Today* 27 (2): 249–60.

Hirsch, Marianne. 1997. *Family Frames: Photography, Narrative, and Postmemory*. Cambridge: Harvard University Press.

———. 2008. "The Generation of Postmemory." *Poetics Today* 29 (1): 103–28.

Koopman, Emy. 2010. "Reading the Suffering of Others: The Ethical Possibilities of 'Emphatic Unsettlement.'" *Journal of Literary Theory* 4 (2): 235–52.

Luckhurst, Roger. 2008. *The Trauma Question*. London: Routledge.

McGlothlin, Erin. 2006. *Second Generation Holocaust Literature: Legacies of Survival and Perpetration*. Rochester: Camden House.

Reiter, Andrea. 2005. *Narrating the Holocaust*. Translated by Patrick Camiller. London: Continuum.

Rosenbaum, Thane. 1999. *Second Hand Smoke: A Novel*. New York: St. Martin's.

———. 2006. "The Audacity of Aesthetics: The Post-Holocaust Novel and the Respect for the Dead." *Poetics Today* 27 (2): 489–95.

Rothberg, Michael. 2000. *Traumatic Realism: The Demands of Holocaust Representation*. Minneapolis: University of Minnesota Press.

Schnell, Lisa J. 2004. "Learning How to Tell." *Literature and Medicine* 23 (2): 265–79.

Sicher, Ephraim. 2005. *The Holocaust Novel*. New York: Routledge.

Terrien, Nicole. 2009. "So Many Silent Voices, Which Are Mine? (Jenny Diski)." In *Voices and Silence in the Contemporary Novel in English*. Edited by Vanessa Guignery, 88–98. Newcastle: Cambridge Scholars.

6 Plight versus Right

Trauma and the Process of Recovering and Moving beyond the Past in Zoë Wicomb's *Playing in the Light**

Dolores Herrero

Trauma has often been described by trauma theorists as a persistent and impossible gateway between the walls of the past and the present. Following Freud's arguments in the comparison he established in his seminal work *Moses and Monotheism* between the history of the Jews and the structure of a trauma, Cathy Caruth relies on the concepts of "latency" and "belatedness" in her well-known volume *Trauma: Explorations in Memory* to define trauma as the successive movement from an event to its repression to its return (1995, 6–7). Caruth explains how traumatic experiences produce:

> A response, sometimes delayed, to an overwhelming event or events, which takes the form of repeated, intrusive hallucinations, dreams, thoughts or behaviours stemming from the event, along with numbing that may have begun during or after the experience, and possibly also increased arousal to (and avoidance of) stimuli recalling the event. (4)

Caruth makes it clear that the event is not assimilated or experienced fully at the time, but only later on, in its repeated possession and haunting of the one who experiences it. John Brenkman's notion of "retrodetermination" which, in Greg Forter's words, presents trauma as "an effect of the interplay between two moments, the second of which retrospectively determines the meaning of the first" (2007, 264), also corroborates this idea. Past traumatic experiences can have such a firm and persistent hold upon the present that this has often led critics to reach the pessimistic conclusion that trauma can be seldom overcome. The disturbing open ending of Zoë Wicomb's novel *Playing in the Light* seems to give some credence to such discouraging perspectives.

Set in post-apartheid Cape Town, *Playing in the Light* describes a traumatised country whose incipient democracy is neither fully reconciled to the past nor confident as to how to face up to the future. Wicomb's novel tells the story of Marion Campbell, a young woman of Afrikaner background who, believing herself to be white, tries to disclose her family history and come to terms with a childhood spent shrouded in an unexplained and suffocating silence. Her quest runs parallel to, and is part of, a new

South Africa striving to emerge from decades of violence, oppression and deliberate blindness. The novel continues to address central concerns of Wicomb's earlier fiction, namely, how apartheid has damaged relationships from a personal to a national level, the conflict between races and generations, where the racist complicity of an older generation is addressed and revisited from the confused point of view of their descendants, and the rigid and suffocating set of binaries upon which the whole of South African apartheid society was established. As Elsie, Marion's father's sister declares, "those were bad times [. . .] this was a place of black and white, not a place of fairness, no room for concessions" (Wicomb 2006, 172).[1] In Marion's world, the burden of the nation's past weighs so heavily that no one wants to talk openly about it, least of all John, her ageing and widowed father, who cannot help feeling guilty for having complied with his wife's wish that they should 'play white,' that is, pass themselves off for white during the apartheid years so as to protect themselves and their daughter from being seen to be the wrong colour. The title of the novel significantly echoes Toni Morrison's contemporary classic *Playing in the Dark*, a short book with an introduction and three essays on the role of race, and specifically Africans and blacks, in American literature. The subtitle is *Whiteness and the Literary Imagination*, and this is what Morrison is mainly interested in because, as she sees it, "it may be possible to discover, through a close look at literary 'blackness,' the nature—even the cause—of literary 'whiteness'" (Morrison 1992, 9). As Morrison goes on to argue, "a real or fabricated Africanist presence was crucial to their sense of Americanness" (6) and also "to any understanding of [their] national literature" (5). "American means white" (47), she concludes, and the black presence only corroborates this and the whites' superiority.

> Africanism is the vehicle by which the American self knows itself as not enslaved, but free; not repulsive, but desirable; not helpless, but licensed and powerful; not history-less, but historical; not damned, but innocent; not a blind accident of evolution, but a progressive fulfilment of destiny. (52)

The title of Wicomb's novel undoubtedly invites the reader to establish some interesting connections with Morrison's contention. Although, as Morrison states, "to identify someone as South African is to say very little; we need the adjective 'white' or 'black' or 'coloured' to make our meaning clear" (47), it is nonetheless true that whiteness was also the norm in South Africa. To put it differently, whiteness was the authoritative standard, the model of correctness that served to regulate action and judgement. As can be read in Wicomb's novel, in the apartheid system "whiteness [. . .] is not a category for investigation" (2006, 120). It is this conviction that leads Helen, Marion's mother, to conclude that "[w]hiteness is without restrictions. It has the fluidity of milk; its glow is far-reaching" (151), and thus reject her 'coloured' condition in order to try desperately to pass herself off for white. Helen's reaction, as Zöe Wicomb argued when discussing in one of her critical essays the problematic condition of the coloured in post-apartheid South Africa, clearly speaks to the persistence of shame in the formulation of coloured identity (1998, 91–107).[2] Helen was prepared to sacrifice all for the sake of being accepted as white, but this alienated John, who was all of a sudden prevented from being in contact with his extended and loving coloured family, with the result that "[t]hey [were] alone in the world, a small new island of whiteness" (Wicomb 2006, 152). In her pursuit of whiteness, Helen takes it upon herself to do away with anything that can give away their past, that is, she is determined to defy history, because "building a new life means doing so from scratch, keeping a pristine house, without clutter, without objects that clamour to tell of a past, without the eloquence—no, the garrulousness—of history" (152). Thanks to Helen's various stratagems (among other things, Helen tells Councillor Carter that her husband lost his birth certificate, and is anxious "to show willing" (143) in order to get him to produce an affidavit that they have always been members of the white community) Marion's parents, once coloured, were reclassified as whites during the Population Registration Amendment Act of 1962.[3] Not only did Helen's decision estrange them from their own families, with all the negative consequences that this had for their own psychic well-being, but it also undermined their relationship with each other and, ultimately, with their only child.

Marion's descent into her family's obscure past is set against the backdrop of the Truth and Reconciliation Commission (TRC), set up in 1995 and chaired by Bishop Desmond Tutu to inquire into the hidden abuses of apartheid and the repressed history of a nation, at which more than 22,000 people testified.[4] The parallel that the novel invites readers to draw is fairly obvious: like the Truth and Reconciliation Commission, Marion feels the need, to use Derek Attridge and Rosemary Jolly's words, to bear "the double responsibility of exposure and acceptance." In other words, she realises she must remember and "narrativize the past in such a way that the future becomes—unlike the past—bearable" (Attridge and Jolly 1998, 3), and a "New South Africa" is somehow rendered possible. The phrase "the New South Africa," used by several characters in Wicomb's novel, was coined by F. W. de Klerk in his speech on 2 February 1990 which proclaimed the end of apartheid, announced Mandela's release from prison, and promised the repeal of the apartheid laws. However, as *Playing in the Light* clearly shows, its conception was anything but smooth and unproblematic. As Dennis Walder points out, "despite the overwhelming endorsement of the reform process by an all-white referendum, the feelings of all races remained at best mixed about the 'New South Africa'" (1998, 153). Contrary to the optimistic exclamation of the librarian who helps Marion that "thank God it's all over, all ancient history, and we needn't bother our heads about that nonsense any more" (Wicomb 2006, 122), the country continued to be rent by resentment and tensions of all kinds, which affected all racial and social groups without exceptions.

As the political situation drastically changes and the balance of power shifts, democracy turns out to be a disorienting experience for the formerly privileged white and affluent classes, who must now "don [their] kid gloves" and "tread gingerly in the New South Africa" (25), while confronting widespread violence and increasing crime rates. To give but one example, Marion's father feels so insecure in the post-apartheid nation that he goes as far as to exclaim in one of his racist fits: "these kaffirs of the New South Africa kill you just like that, just for the fun of it" (13). Marion will reach a similar conclusion when two toughs demand money for protecting her car in her own parking bay: "You can't go anywhere nowadays without a flock of unsavoury people crowding around you, making demands, trying to make you feel guilty for being white and hardworking, earning your living; and of course there is no getting around it" (28). In the new Cape Town, the colossal gap between the rich and the poor makes Marion, who regards herself as belonging to neither of those two groups, feel terribly uncomfortable:

> Marion cannot accuse the opportunistic layabouts of Cape Town of being late risers, of suffering from Monday blues. The streets are already dotted with ragged people wiping the sleep out of their eyes, buffing their begging bowls, gearing up to bully and abuse the law-abiding citizens who will not be taken in by them. She is equally impatient with the idle rich, women of leisure who, like people of the townships, stretch out in the sun. (25)

Marion's life was that of a white child, albeit from the poorer side of town, cramped with modest houses lacking the screened verandahed steps which allowed the more affluent whites living above the Main Road "to spend all day outside in the ambiguous space between private house and public street" (9). Unlike these rather more privileged whites, Marion lived in a terraced house very close to the pavement, and was forced by her mother to be confined, to keep indoors and, most important of all, to keep out of the sun, even in summer. Being neither rich, nor poor, Marion's family straddles both worlds. Significantly enough, the expensive flat in which she now lives, a mature woman and owner of a successful fancy travel agency, also corroborates Marion's vulnerable liminality. The opening sentences of the novel speak for themselves:

> It is on the balcony, the space both inside and out where she spends much of her time at home, that it happens. A bird, a speckled guinea fowl, comes flying at a dangerous angle, just missing the wall, and falls dead with a thud at Marion's feet. [. . .] There is silence overhead. Will the others, the enemies, line up on her balcony wall to pay their respects? Should she withdraw? (1)

The setting of Marion's current flat, and more particularly her balcony, highlight themes of class aspiration and conflict, safety, lack of homeliness and, most important of all, fear and psychological uncertainty. In her critical essay entitled "Setting, Intertextuality, and the Resurrection of the Postcolonial Author," Wicomb argues that setting is not simply a fictional location, since it can also provide precious information on the psychological condition of characters and the socio-political situation of the society that forms the context of the work. To quote her own words:

> setting is the representation of physical surroundings that is crucially bound up with a culture and its dominant ideologies, providing ready-made, recognizable meanings. In other words, setting functions much like intertextuality. (2005, 146)

The balcony is clearly referenced as the pseudo-interior/pseudo-exterior space that she prefers. However, it is there that the bird falls dead, which immediately arouses Marion's feelings of helplessness and anxiety. The introduction to this liminal space, therefore, initiates the narrative's preoccupation with Marion's hidden traumatic experience of hybridity and disorientation, which will lead her to become a control freak—"All that matters," she often says to herself, "is that things are returned to their rightful places" (Wicomb 2006, 16)—and look for some safe shelter within her own luxurious, but ultimately disturbing, flat. The big four-poster bed where she sleeps is much more than a marker of her success; it is her private hide-out,

> a house in itself, into which she can retreat from the larger one when she needs the cocoon of draped muslin after a hard day's work, the noise of the world dampened to a distant hum. [. . .] a bed that was hardly an item of furniture. Rather a bower for an egte fairy princess, who would lie for a hundred chaste years in gauzed limbo, waiting for the world to change into a better, a more hospitable place. (2)

However, no matter how hard she tries to hide, she often suffers sudden fits of panic that threaten to do away with her precarious balance and sense of self. Her bed, so far envisioned as her shelter, uncannily becomes her prison-house:

> But lately, the four-poster bed has turned against her. There have been times [. . .] when something buzzes in her ears, a sense of swarming that grows louder and louder [. . .]. Then, for a moment, she seems to gag on metres of muslin, ensnared in the fabric that wraps itself round and round her into a shroud from which she struggles to escape. (2)

From the very beginning of the novel, it becomes obvious that her turbulent family story has left her with no clear identity. Her light hair, pale complexion, and relatively privileged upbringing never led her to question her race. However, her father's evasive silence, together with her unaffectionate mother's obsession with keeping oneself to oneself and preventing any relatives and friends from visiting them, cast surreptitious doubts upon their social, and racial, position. Marion's name undoubtedly points to this in-betweenness. Her father, "the young man from the Karoo, who loved the sea" (22), wanted to call her Marina, seeing that he could not call her Mermaid, his truly favourite name. In keeping with her white aspirations, Helen bluntly rejected such a name: "Marina, she scoffed, that's not a name. Only hotnos give their children stupid names like that" (125). Although in the end they settled for Marion, in private John kept on calling her Marina, Marientjie, his darling *meermin* (mermaid in Afrikaans), which clearly suggests her hybrid condition. Suspended in between two irreconcilable realms, Marion belongs in a state of indeterminacy that her mother cannot possibly accept, and for which she blames her pusillanimous husband: "No good being half woman and half fish, half this and half that; you have to be fully one thing or another, otherwise you're lost. Mermaids are the silly invention of men who don't want to face up to reality, to their responsibilities, the fantasy of losers who need an excuse" (47). Like the mermaid in the fairytale her father once told her, "who gets mixed up with humans and who may well have come to a sorry end" (46), Marion has always found it difficult to move around, communicate, and make friends. At one point in the novel she is accused by Johan, a boyfriend with whom she had a short and problematic relationship, of being "cold-blooded, a plain cold fish" (22). Being the owner of a successful travel agency, Marion paradoxically has "an aversion to travel" (40), which she regards as a futile attempt to escape reality, to "[kid oneself] that it is possible to get away" (39). Another reason why Marion dislikes travelling so much is her absolute incapacity to appreciate and interact with the landscape around her, a dislike which clearly underpins her confused and traumatised identity, and by extension her problematic relationship with her homeland: "For Marion, the truth is that she is passing over the land, her only contact being through the cushioned rubber tyres of her moving vehicle" (80). Nor is she interested in reading fiction, for similar reasons.

As Cathy Caruth argues (1995, 4–10), it is the very unassimilated nature of trauma, that fact that the event was neither acknowledged nor experienced fully at the time, that later returns to haunt the survivor. When Marion was only five, Tokkie, the black servant who came once a week to see them and keep an eye on things in their house in Observatory, the old woman who looked after her, "loved her, spoiled her rotten" (Wicomb 2006, 32), died. This will be one of the traumas which, together with that of having an unloving mother, will haunt Marion for years, rendering her unable to enjoy healthy relationships of any kind, and making her dream of the same place once and again. This recurrent place is envisioned as yet another liminal space, since, as Marion says, "the focus is clearly on doors"

(31): a house in a green valley, a loft with its black wooden door, against which a wooden ladder leans, and an old black woman sitting inside.

> Marion's is a recurring dream, even if there are minor variations. [. . .] This time, all the doors and windows are shut; the woodwork is painted black. When she looks up, the loft door bangs, although there is no wind. She climbs up the ladder. Her mother is at the foot of the ladder, pleading with her to come down, giving the ladder a gentle shake to frighten her. But Marion carries on [. . .] When her head reaches the height of the loft door, she pushes it wide open so that a broad shaft of light falls across the floor. An old woman sitting on a low stool is illuminated; the light falls on a white enamel basin on her lap. Her face, sunburnt and cracked like tree bark, is framed by the starched brim of a white bonnet. [. . .] The old woman is busy, does not look at her. [. . .] Marion waits to be invited into the loft, but eventually gives up and starts going down the ladder, one foot guardedly following another in a backward descent so that the old woman disappears slowly in bands of darkness. (30–31)

Marion lost the person who loved her most. Paralyzed and unable to face up to such loss, she can only hide and grope for her in her dreams. Her reaction is, therefore, melancholic. As Freud explains in his seminal essay "Mourning and Melancholia" (1917), whereas mourning usually implies the death of the lost object, melancholia may appear under any circumstance whereby the object loses its status as a loved one, that is, when the object "has been lost as an object for love" (Freud 2001, 245). The conscious nature of mourning facilitates the process of withdrawing the libido's attachment to that object. On the contrary, the strong unconscious fixation between the ego and the lost object that melancholia entails turns this process into a rather more complex and pathological condition, which often derives into identification and love-hate ambivalence. Dominick LaCapra (2001), yet another well-known trauma theorist, also distinguishes between mourning and melancholia, and relates both concepts to the processes of "working through" and "acting-out" (66) respectively. In the case of melancholia, the traumatised subject remains trapped in a compulsive search for the beloved object and the re-enactment of the original trauma. In the case of mourning, on the other hand, the subject is able to specify and gain some distance from her/his traumatic loss. Abraham and Torok (1994) also offer an interesting reconceptualisation of Freud's notions. Rather than distinguishing between mourning and melancholia, they talk about introjection and incorporation. Both critics define introjection as "the process of broadening the ego" (127). As in mourning, introjection allows the subject to refashion itself, channel the new account, pain, desires, and thus accommodate them within her/his situation (110). In contrast, incorporation is "the refusal to acknowledge the full import of the loss, a loss that, if recognised as such, would effectively transform us" (127). However, unlike Freud, who put great emphasis on disappointment as the main cause of melancholia and the subsequent ambivalent love-hate relationship between subject and lost object, Abraham and Torok claim that, for the melancholic, the lost object keeps its idyllic status, since the main causes of the loss are taken to be external. Tokkie's loss turned Marion into a melancholic subject, haunted by dreams that enshrouded her grief and fears. However, pace Freud and Abraham and Torok's theories, in Marion's psyche Tokkie turns out to be an ambivalent figure: although she preserves her idyllic halo, she forsakes Marion by refusing to look at her.

As regards Marion's mother's death, it could be argued that this trauma, to use LaCapra's terms again, did not originate in a loss, but rather in an absence, an abstract entity that cannot be specified and symbolised because the subject was never actually in possession of it (2001, 49--53). One cannot lose what one never had, but this does not prevent one from feeling the lack all the same. Marion feels that she never had a mother, that her mother never really wanted her. Helen's unexpected pregnancy was riddled with anxiety and, even after Marion was born with pale skin and smooth hair, she experienced no relief, and "thought of the baby as an uninvited guest, arriving with an extraordinarily large, cheap suitcase that bumped along through the birth canal" (Wicomb 2006, 134). When, being only a child, Marion is told off by Helen for rolling half naked in the grass "like a disgusting native" (60), she pleads with her that she wants to have a sister to play with, and this is the bitter dialogue that follows:

> Oh no, you don't want a sister, she said in a horribly quiet voice.
> But I do, the child whined, I want, I want, I want a sister . . .
> Stop it, shut up, Helen shouted brutally, I can't have any children, we don't want any children, so just shut up, and with distaste, with her shoe, she shoved the child's clothes towards her. Marion ran off, crept into bed in her vest and broekies; she had upset her mummy. She was stricken with guilt and remorse. (60–61)

Unable to cope with the guilt and remorse that go hand in hand with her lack and the pain brought about by the loss of Tokkie, her surrogate mother, Marion creates—to avail of Abraham and Torok's ideas again—a secret tomb/crypt within her psyche that "includes the actual or supposed traumas that made introjection impracticable," and where "the objectal correlative of the loss is buried alive as a full-fledged person" (Abraham and Torok 1994, 130). As regards Marion, then, the old house in the dream, and the old black woman sitting inside, are nothing but the materialisation of her own intrapsychic crypt.

Moreover, as is frequent in trauma, another traumatic event suddenly connects with the original event which has remained hidden for a long time in Marion's mind. It is only after she sees in a newspaper the photograph of a coloured woman called Patricia Williams, and reads about her ordeal at the hands of the Security Police, that Marion realises that "there is something secret, something ugly, monstrous, at the heart of their paltry little family" (Wicomb 2006, 58). What she had at first disdainfully regarded as yet "[a]nother TRC story" (49), all of a sudden compels her to step out onto her balcony, where she has the revelation that will unconsciously allow her to make the connection, and thus set about undertaking her family quest.

> From her balcony, she stares in horror at an enlarged face floating on the water, a disfigured face on the undulating waves, swollen with water. [. . .] It is not until she goes back indoors that recognition beats like a wave against the picture window: Tokkie, it is Tokkie's face on the water. (55)

Marion identifies the phantom in the sea with Tokkie. Until that moment, Marion had refused to face up to the source of her life-long numbness. She had never been interested in therapy and had despised those who did, since she couldn't help regarding them as "indulgent, effete, English types, who do not know how to roll up their sleeves and get on with things" (3). There was no point in spending time and money in discovering the obvious, that "she had a peculiar childhood; that her parents loathed each other; that her mother, like all mothers, was responsible for her insecurity" (3). However, this unexpected connection—Patricia Williams bears some undeniable physical resemblance to Tokkie, and by extension to herself— makes Marion's intrapsychic crypt crumble to pieces, which explains why, when staring at the peach she has just peeled, Marion feels "naked, slippery [. . .]. Hurled into the world fully grown, without a skin" (101). This unanticipated revelation also urges Marion to look into her family's history. All of a sudden, she feels the need to do away with the deadly silence and secrecy with which her mother had muffled their lives, and for the first time sees her past as "contained in endless dreary rows of parcelled days, wrapped in tissue paper, each with its drop of poison at the core" (61), in a word, with poisonous secrets that she must unveil. She does not know where this eerie certainty comes from, but is sure that the mystery is related to her own birth. She must have been an adopted child, which would explain why in her mother's face "she often caught a look of naked resentment" (62). Although she can't have been biologically connected to Tokkie, she guesses that this old black lady was party to the adoption, and became in turn her true mother figure. After all, "[i]t was Tokkie who understood the child's loneliness and loss and unease, and so lavished love on her" (62). Patricia Williams' face has alerted her to the truth, has turned into the shamanic figure who commands that she should start off on her painful quest.

Marion's father insists that he does not remember Tokkie's surname. Marion then searches for answers in Helen's Black Magic chocolate box, the box that contains the remains of her mother's possessions, but also in vain, for the box, like Marion's disturbing memories of her mother, gives nothing away. In a desperate attempt to find out the truth, Marion asks Brenda, her only coloured employee at the travel agency, to help her discover what became of Tokkie, apparently because she believes that all coloureds have connections to one another. Interestingly enough, Marion's assumption proves to be correct. Travelling with Brenda, Marion learns, during a Ulysses-like epiphany scene in Wuppertal, that Tokkie was her mother's mother, that is, her biological grandmother, that Patricia Williams is a close relative of hers, and that the house that she recurrently sees in her dreams must be his father's family farm, which she visited with him when she was little. Light and liminal spaces—doors and windows—preside over the whole dream. Once again, an unexpected connection brings about Marion's epiphany, and shock, at her parents' lies, at her having been brought up white and discovering she is nothing but a hybrid. Jennifer L. Apgar brings all these ideas together when she argues:

> [t]he woman in the dream connects in Marion's mind to Tokkie [. . .]. The house itself connects to John's family home remembered from a childhood visit [. . .]. In this dream Marion's subconscious appears to be preoccupied with issues of home. In this context, light is a metaphor for understanding or insight and is controlled by the threshold spaces of doors and windows. (Apgar 2008, 50–51)

It is now that the phantom that Marion saw on the surface of the sea, and that she immediately identified with Tokkie, makes full sense. Abraham and Torok's theory of the phantom as an alternative explanation for cases of pure identification may help to account for this phenomenon. As these critics put it, "[the phantom] works like a ventriloquist, like a stranger within the subject's own mental topography" (1994, 173). Unlike the melancholic subject, who denies and encrypts her/his own wound caused by the loss of an ever-beloved subject (135), the haunted subject becomes her/his phantom's unintentional tomb, which encloses her/his ancestors secrets/traumas (171–72). In other words, "here symptoms do not spring from the individual's own life experiences but from someone else's psychic conflicts, traumas, or secrets" (166). The phantom encapsulates the interpersonal and transgenerational consequences of silence: although "the dead do not return, [. . .] their lives' unfinished business is unconsciously handed down to their descendants," and can "lead a devastating psychic half-life in us" (167).

Marion's grandmother and mother kept a secret to themselves, a secret buried by guilt and shame which they passed on to Marion as a phantom, as transgenerational trauma. However, as Abraham and Torok go on to argue by alluding to Hamlet's phantom, these haunting ghosts should never be taken at face value, since "a ghost returns to haunt with the intent of lying: its would-be revelations are false by nature. [. . .] The secret revealed by Hamlet's "phantom" and which includes a demand for vengeance, is merely a subterfuge" (188–89). Marion initially misperceives the true relationship between the two women. She accuses her mother of having humiliated her grandmother, and wonders: "How am I to bear the fact that my Tokkie, my own grandmother, sat in the backyard drinking coffee from a servant's mug, and that my mother, her own daughter, put that mug in her hands?" (Wicomb 2006, 103). Marion thinks that this arrangement was forced on Tokkie by Helen. It is the narrator who discloses the truth: contrary to what Marion thinks, it was Tokkie who created her own role. "Helen did not think it was necessary, but [Tokkie] insisted that one could not be too careful with the neighbours" (132). This is the secret that both women took to their grave.

Now that she knows her true identity, that she has a past and a family, Marion becomes, for the first time in her life, aware of her "loneliness, this vast emptiness before her" (177). Nothing is the same, "the history of the country, too, has slid from the textbook into the very streets of the city" (177), and Marion feels strong enough to face up to that image on the water, to look that face directly in the eyes. Besides, she is now able to regard Maria, her servant, as "a paper-thin, arthritic old woman [. . .], a frail creature" (177), in short, as a human being as vulnerable as herself. Last but not least, it is now that Marion becomes fully aware that she betrayed her dear friend Annie Boshoff when she was only eight years old. Annie and her family had to leave the community because her father was "caught with his pants down on top of a coloured girl" (193), and was accordingly accused of immoral behaviour. Before departing, Annie gave Marion their shared scrapbook, but Marion was so afraid of being related to her that she dropped the scrapbook into the dustbin in the backyard and decided to forget about her friend altogether. Her painful revelation is that she was no better than her parents.

In addition, and much to the reader's surprise, Marion takes the unexpected decision to travel to Europe. She visits emblematic European cities like Berlin and London. However, these cannot provide her with any comfort. Marion can't help feeling "in the wrong hemisphere," she "is invaded by the virus of loneliness," and realises how dull and empty the words "Love you," which the English say to close every single conversation, can be (188). The most significant episode of this adventure will undoubtedly be her encounter with Dougie, the elderly Scotsman who insists on telling her the complicated history of the Campbells, Marion's supposed ancestors. This, together with the fact that, before Marion departs, Dougie buys a haggis for her and a tie in Campbell tartan for her dad, seems to corroborate Marion's father's fantasy of their truly European pedigree. On the other hand, Dougie's allusion to the Scottish water spirit, the kelpie, which takes particular delight in teasing travellers, can conversely be interpreted as a warning, as yet another unreliable phantom that counters and undermines the previous optimistic assumption. During this journey, she also starts reading novels, in particular such politically-laden South African novels as *The Conservationist*, by Nadine Gordimer, and *In the Heart of the Country*, by Coetzee. The fact that these novels have already been studied as 'trauma novels' suggests that some kind of intertextuality was intended by the author. Although the novel does not seem to explore these connections in depth, it is nonetheless clear that Marion starts to make connections between what she reads and her own present reality, her past, and her convoluted family history. All of a sudden, she realises the true nature and implications of reading: "reading is, or should be [. . .] absorbing words that take root, that mate with your own thoughts and multiply" (190). It is now that she becomes aware of "the hole in her chest" (190), but nonetheless decides to go on reading and living through her pain and crying in order to get to know those dark decades when her family, the Campbells, were playing in the light. The rectangle of light that Marion sees opposite the window, which constantly changes shapes and keeps on appearing and disappearing, is nothing but the ultimate projection/embodiment of her family's turbulent past and uncertain identity.

However, what might have facilitated her coming to terms with her past, and hence with her present, turns out to have been a distraction, a mere parenthesis at the end of which Marion resumes her former emotionally crippled life, without having been able to overcome the trauma that has been haunting her for years. Everything is changing too fast, and the impossibility of knowing what the New South Africa will bring them all in the near future prevents Marion from shaking off her ingrained fears and uncertainties, and she becomes instead obsessed with the questions that tortured her in the past, namely, the need to distinguish between what things are and were, and what they mean and meant.

> How can things be the same, and yet be different? Is the emptiness about being drained of the old, about making room for the new? Perhaps it's a question of time, the arrival of a moment when you cross a boundary and say: Once I was white, now I am coloured. If everything from now on will be different (which is also to say the same), will the past be different too? (106)

Marion's inconclusive and disappointing ending could be said to be in tune with those offered by many late-apartheid novels, which, according to Elleke Boehmer, "could not in most cases be taken as other than a closing down or narrowing of possibility," since they give us, as in Marion's case, "escapes, but without clear destinations, departures which are headed for culs-de-sac, caught in a void" (Boehmer 1998, 45). There is in Marion's elusive final attitude a reluctance to give any sort of positive reading about what might happen in South Africa from now on. Marion's ending is, to quote Boehmer's words again, "arrested in a difficult and frozen now" (48). Soon after her arrival in Cape Town, Marion learns that her coloured friend Brenda has looked after her father and made friends with him during her

European tour, and is actually writing the story of his life. Her hostile reaction towards Brenda, the "agent of reconciliation" (Wicomb 2006, 214), who was kind enough to organise a surprise welcome party for Marion, in which all of her relatives could at last get together, comes as a shock. Marion simply cannot take Brenda's act of writing, and vehemently tells her friend to get out of her flat.

There is yet another relevant episode that might help to account for these two women's different final reactions. One day, as they were approaching Wuppertal in their search for Tokkie, their car ran into Outa Blinkoog, a man in harness, dragging behind him a ramshackle cart full of objects made of tin. Outa Blinkoog is a nomad, a free spirit who believes that a person must "do as he pleases, go where and when he likes" (90). He will not let them go without a present, and gives them a lantern that they both must share. This is no ordinary lantern, since its "last hour of candlelight is sweetened with bright colour, so there is no place for sadness" (91). Significantly enough, it was Brenda who began writing Marion's father's story, "the story that [she] wanted to write, the story that should be written [. . .] with his pale skin as capital, ripe for investment . . . " (217–18), thanks to the help of the lantern—Brenda, the truly coloured woman who has suffered in her own flesh the lethal consequences of apartheid, who has internalised the negative images that this castrating system has ruthlessly imprinted on her black mind to the point of making her hate, not only her own home, but also "the sponginess [. . .] the smell and the softness" of her affectionate old mother's body (65). The lantern's inspiring light managed not only to drown out the noise in her poor overcrowded house, thus providing Brenda with the peace and quiet she needed to start writing, but also to illuminate and invite Brenda to open herself up to the other so as to develop new ethical modes of cross-racial interchange and understanding. Brenda consequently becomes the listener who, according to Shoshana Felman and Dori Laub, bears witness to Marion's father's traumatic testimony, to his "black hole both of knowledge and of words" (Felman and Laub 1992, 65). She turns into his "companion" (59), and remains attentive to his plight so as to help him to raise his repressed predicament into consciousness. Yet, Brenda is not dependent on the lantern's light, since she can, according to Gail Jones, "speak shadows," which implies

> an admission of uncertainty, a calculation of difficulty, and awareness that justice—and human relations—is rarely written in black and white. It requires commitment to some state of thinking which radically oscillates across time, between past and future, and is therefore a condition of process and hope, rather than of certainty. (Jones 2011, 6)

Brenda has become aware of the fact that the binary forms with which we are taught to think are never a sufficient—or fair—basis for a moral life, and is brave enough to dare to transcend the silences imposed by the simple oppositions and binarities that the apartheid system enforced. Brenda is therefore conscious of the urgent need, as André Brink put it, to "imagine the real" (1998, 24), that is, to face up to new testimonies which, disturbing as they may be, can nonetheless allow the citizens of the new nation to grow and mature by coming to terms with the dark places—the silences—in themselves. Brenda's need to tell the underside of apartheid history, and to bring to the fore its implications for the present and future, goes hand in hand with her desire to find a new form of narration, capable of acknowledging difference, and lights and shadows alike, without demonising them, and without fetishising them either. Unlike Marion's disappointing ending or "frozen now," then, Brenda's final courage and generosity can be said to illustrate the new kind of open ending that, according to Elleke Boehmer, was much wished for in post-apartheid South African novels. In the case of Brenda, *Playing in the Light* offers an ending that "allow[s] for new beginnings, for gestative mystery, the moments and movements following apocalypse, also the dramatization of different kinds of generation and continuity" (Boehmer 1998, 51). The novel's plural and ambivalent conclusion therefore stands for an open-endedness that allows for multiple possibilities and makes room for new and various ways of thinking about the future in contemporary South Africa. After all, as Marion says, "that is what the new is all about—an era of unremitting crossings" (Wicomb 2006, 107). *Playing in the Light* offers, to use Rob Nixon's words, "an approach to differences that breaks with smiling multiculturalism and its ugly mirror image, apartheid, by recognizing that inequalities in power slice across the sites of identity" (Nixon 1991, 32).

In contrast to Brenda, who knows how to benefit from the light of Outa Blinkoog's extraordinary lantern without becoming dependent on it, Marion, the privileged coloured woman who was brought up as if she were white, demands that she should have the lantern back, but is unable to accept and make the most of her father's fascinating life story. Marion has been longing to know the story of her family, but now that she has disclosed it and somebody else is willing to narrate it and bring it to light, she desperately tries to silence it, to keep it hidden, thus making it clear that, in spite of all her efforts, she has not managed to forgive her parents and work through her personal trauma. Marion has failed to understand that, as the Italian philosopher Giorgio Agamben states, "justice is not about repentance, so much as it is about recovering one's own and others' possibilities and potentialities—a kind of plenitude for every life, one that requires imagining backwards (to regret historical mistakes), and forwards, to constitute a more-just future" (in Jones 2011, 3). If her mother strove to play in the light and Brenda has learnt to speak shadows, it seems that, so far, Marion can unfortunately only keep quiet in the dark.

*The research carried out for the writing of this essay is part of a project financed by the Spanish Ministry of Economy and Competitiveness (MINECO) (code FFI2012–32719). The author is also grateful for the support of the Government of Aragón and the European Social Fund (ESF) (code H05).

NOTES

1. From *Playing in the Light* by Zoë Wicomb, copyright © 2006 by Zoë Wicomb. Used by courtesy of Umuzi, an imprint of Random House (Pty) Ltd.
2. In this essay, Wicomb states that the recent history of coloured politics in South Africa does away with Homi Bhabha's association of a subversive hybridity expressed specifically in terms of biological metaphor. The fact that with the post-apartheid vote coloureds chose to align themselves with the showy multiculturalism of de Klerk's Nationalists, Wicomb claims, points to their deeply-rooted feeling of shame. In this context, the function of the recognition of difference in the subject classified as hybrid is nothing but self-abnegation. The fetishisation of difference, which goes hand in hand with the stigmatisation of race, always produces negative effects. Neither coloured shame nor white guilt can contribute to laying the foundations of a healthy and solid post-apartheid era, Wicomb concludes.
3. The Population Registration Amendment Act of 1962 allowed South African subjects to apply for a racial classification other than that given them as a result of the enforcement of The Population Registration Act of 1950, which required that each inhabitant of South Africa should be classified and registered in accordance with their racial characteristics as part of the apartheid system. Social rights, political rights, educational opportunities, and economic status were largely determined by which group an individual belonged to. There were three basic racial classifications under the law: black, white and coloured (mixed). Indians (that is, South Asians from the former British India, and their descendants) was later added as a separate classification as they were seen as having no historical right to the country. An Office for Race Classification was set up to overview the classification process. Classification into groups was carried out using criteria such as outer appearance, general acceptance and social standing. Because some aspects of the profile were of a socio-economic nature, several Amendment Acts were passed to allow for reclassifications, and a board was established to conduct that process. This law worked together with other laws passed as part of the apartheid system, such as the Prohibition of Mixed Marriages Act of 1948, whereby marrying a person of a different race was illegal, and the Immorality Amendment Act (Immorality Act) of 1957, which regarded any attempt to conduct a relationship with a member of a different race as a crime. Although the South African Parliament repealed these Acts in 1991, the racial categories defined in them remain ingrained in South African culture and they still form the basis of some official policies aimed at correcting past economic imbalances. For more information on the subject, see Leonard Thompson (2001, 182–214).
4. As is well known, the TRC offered a forum for victims to bear witness against the apartheid state and its agents, and offered amnesty to those who confessed their implication in state-perpetrated crimes, except in cases of extreme sadism or those that manifested personal revenge. The whole process was, nonetheless, highly controversial for, more often than not, there

was no way of avoiding errors of judgment in such a difficult context, and the offer of amnesty was fairly polemical in too many cases.

WORKS CITED

Abraham, Nicolas, and Maria Torok. 1994. *The Shell and the Kernel: Renewals of Psychoanalysis*. Vol. 1. Edited and translated by Nicholas T. Rand. Chicago: University of Chicago Press.

Apgar, Jennifer L. 2008. *Performing Passing: Theatricality in Zoë Wicomb's Playing in the Light and Nella Larsen's* Passing. MA Diss., College of Arts and Sciences, Georgia State University.

Attridge, Derek, and Rosemary Jolly. 1998. "Introduction." In Derek Attridge and Rosemary Jolly, 1–13.

———, eds. 1998. *Writing South Africa: Literature, Apartheid, and Democracy, 1970–1995*. Cambridge: Cambridge University Press.

Boehmer, Elleke. 1998. "Endings and New Beginning: South African Fiction in Transition." In Derek Attridge and Rosemary Jolly, 43–56.

Brink, André. 1998. "Interrogating Silence: New Possibilities Faced by South African Literature." In Derek Attridge and Rosemary Jolly, 14–28.

Caruth, Cathy, ed. 1995. *Trauma. Explorations in Memory*. Baltimore and London: Johns Hopkins University Press.

Felman, Shoshana, and Dori Laub. 1992. *Testimony: Crisis of Witnessing in Literature, Psychoanalysis, and History*. New York and London: Routledge.

Forter, Greg. 2007. "Freud, Faulkner, Caruth: Trauma and the Politics of Literary Form." *NARRATIVE* 15 (3): 259–85.

Freud, Sigmund. (1917) 2001. "Mourning and Melancholia." In *On the History of the Psycho-Analytic Movement, Papers on Metapsychology and Other Works. The Standard Edition of the Complete Psychological Works of Sigmund Freud*. Vol. 14. Edited and translated by James Strachey, 243–60. London: Vintage.

Jones, Gail. (2008) 2011. "Speaking Shadows: Justice and the Poetic." Accessed March 16 2011. http://www.tru.ca/cicac/readings/jones-speakingshadows.pdf.

LaCapra, Dominick. 2001. *Writing History, Writing Trauma*. Baltimore, MD: Johns Hopkins University Press.

Morrison, Toni. 1992. *Playing in the Dark: Whiteness and the Literary Imagination*. Cambridge, MA: Harvard University Press.

Nixon, Rob. 1991. "'An Everybody Claim Dem Democratic': Notes on the 'New' South Africa." *Transition* 54: 20–35.

Thompson, Leonard. 2001. *A History of South Africa*. Johannesburg and Cape Town: Jonathan Ball.

Walder, Dennis. 1998. *Post-Colonial Literatures in English: History, Language, Theory*. Malden, MA: Blackwell.

Wicomb, Zoë. 1998. "Shame and Identity in the Case of the Coloured in South Africa." In Derek Attridge and Rosemary Jolly, 91–107.

———. 2005. "Setting, Intertextuality and the Resurrection of the Postcolonial Author." *Journal of Postcolonial Writing* 41 (1): 144–55.

———. 2006. *Playing in the Light*. Johannesburg: Umuzi.

7 Seeing It Twice
Trauma and Resilience in the Narrative of Janette Turner Hospital[*]

Isabel Fraile Murlanch

Most trauma theorists have described the nature of trauma as a certain kind of interaction between the past and the present. One only has to remember Freud's concept of *Nachträglichkeit*, which John Brenkman rendered into English as "retrodetermination," and which, as explained by Greg Forter, presents trauma as being "less a matter of punctual events intruding upon an unprepared psyche than an effect of the interplay between two moments, the second of which retrospectively determines the meaning of the first" (2007, 264). Another case in point is Cathy Caruth's concept of "belatedness," or her famous definition of trauma as a "possession by the past" (1995, 51).

Such is the hold which trauma is supposed to have over the present that it often seems that there is but little hope of overcoming the trauma. Janette Turner Hospital's narrative sometimes seems to accord with such bleak perspectives, as in the following exchange between a man who was a prisoner of the Japanese during the Second World War and the interviewer, Catherine Reed:

> "How long does it take, then, for the worst after-effects of the trauma to wear off, would you say?"
>
> Mr Kenney, startled, looks directly at the point where it can be assumed Catherine Reed is sitting, off camera. He raises his eyebrows in a kind of shock, as though she has asked: And when do day and night stop arriving? When do birth and death disappear? Then he looks away again. (Turner Hospital 1992, 125)

Nonetheless, in contrast to the position held by those who argue that there is no way out of trauma, other theorists (probably, most famously, Boris Cyrulnik) have seen this interplay as opening up the possibility of generating resilience. The term resilience originally meant simply the resistance of a material which allows it to recover its original shape after having been bent, smashed, etc, and it was later transposed to the psychological sphere to express people's capacity to recover from trauma. Its main ingredients, Cyrulnik says, are suppleness, dynamism, resourcefulness and good humour: a mixture which generates self-reliance and joy (2001, 212–13). Now, according to Cyrulnik, one of the main factors that makes it possible for resilience to develop is the way in which present and past intertwine in the narratives the wounded person builds up to make sense of the aggression s/he has suffered. Cyrulnik, like Freud, considers the narrative of the aggression to be of vital importance but, unlike Freud, he holds that this second moment, this second aggression delivered by the representation of the blow (2001, 182),[1] need not lead inexorably to trauma, but may also open up the possibility of resilience. Facing an event that would traumatise other people, those who are truly resilient may feel wounded, but not traumatised (184). Resilience, in fact, depends largely on one's own ability "to organise one's own history" (214), so that representation may turn out to be healing as well as traumatic. As Cyrulnik expresses it, the answer to one of the two basic questions that resilience tries to answer, "How did I manage to survive?" is to be found in research into "personal, family and social narratives," which explain "how we can modify our representations" (2005, 258). Does this mean that trauma can actually be completely cured? Cyrulnik is conclusive in this respect: "one must be clear: there is no possibility of reversing a trauma [. . .]. A precocious wound or a serious emotional commotion leave brain and affective traces that remain hidden once development has been resumed. The sweater thus woven [. . .] may become beautiful and warm again, but it will always be different" (2001, 124). Cyrulnik concludes by confidently asserting that, even if trauma is not reversible, it can always be repaired, in such a way that it can mark a person's life forever without necessarily leading her/him to neurosis (161).

The tension between these opposite beliefs, that trauma lasts forever, and that new paths may be created by resilience, is mirrored by Turner Hospital too at the very beginning of her novel *Charades* (1988). Whereas she opens it with the following quote from Primo Levi: "Once again it must be observed, mournfully, that the injury cannot be healed: it extends through time," she nonetheless provides it with the following dedication: "For my father and mother who taught me that love is rich and redemptive whatever costumes and guises it wears." What Turner Hospital is supporting here is nothing less than Boris Cyrulnik's conviction that love, together with narrative, is one of the ways in which trauma can be overcome. Not that she is too optimistic in this respect. When expressing her interest "in how people negotiate the rest of their lives in the wake of catastrophic events," she highlights the fact that

> There seem to be three possible responses: (a) there are those who never recover: they become suicides, or are institutionalized; (b) there are those who survive but in damaged and dysfunctional ways of greater or lesser severity (i.e., they become alcoholics; or they muddle through relationships and jobs as losers; or they become agoraphobic, or mute);

and (c) a few will emerge strengthened, more compassionate and humane, triumphant. (quoted in Greiner 2007, 338)

Even if not many achieve, from her point of view, resilience, she has always remained fascinated by "the secret of this third category," which, she has, she says, "always been exploring" (338). In the light of some of the auto-biographical details she has offered in different interviews, it would seem that she considers herself one of the happy few who are able to leave trauma behind by means of the power of narrative. She, for example, refers to her mystery novel *A Very Proper Death* (published under the pseudonym Alex Juniper because she did not want it to be "confused with her literary novels") as an exercise of "exorcism of trauma" (1990, 338) carried out in order to free herself from the weight of having been mugged by four youths in 1987. In the same interview, she also refers to her first four novels as "my personal immunisation programme" (325). At other times, however, she seems to think that overcoming traumatic events is a matter of mere chance, that neither love nor narrative have anything to do with it. In an interview with Selina Samuels (2006, 156), she declared: "I'm constantly writing about pairs of women in my novels: one goes under and one doesn't. The reason I didn't go under I think was just by sheer good luck." In any case, what seems undeniable is that she does believe resilience to be possible. And this possibility becomes, more and more often in her work, stronger, perhaps because, in David Callahan's words, "hope, however banal it sounds, is a fierce imperative in Hospital's words" (2009, 204).

As for Cyrulnik, his firm conviction that resilience is, indeed, possible, is not a mere belief either, but, as in the case of Turner Hospital, the result of his own experience. A member of a family of migrant Russian Jews, he managed, at the age of six, to run away from a concentration camp from which the rest of his family never returned. In spite of the fact that he was not able to attend school until he was eleven, with the passing of time he became a neurologist, psychiatrist and psychoanalyst, as well as one of the founders of human ethology and the leading figure of resilience studies in France.

His theory of resilience is therefore, both the product of personal experience, and the result of many years of serious research, which have crystallised in over half a dozen books. It is, again, in *Talking of Love*, that Cyrulnik formulates the second basic question resilience tries to answer: "How is it possible to hope when there is no hope?" Studies of attachment bonds, he adds, have already provided many important clues (2005, 258). Here, in fact, Cyrulnik is drawing on John Bowlby's 1979 seminal work, *The Making and Breaking of Affectional Bonds*, which affirms that a person's resilience depends basically on the structure of the interpersonal links an individual has established during her/his early childhood: to put it another way, resilience depends on whether or not a person has been loved, cared for, and provided with a sense that the world is a good, safe place to live in.

In what follows, it is my intention to explore the ways in which resilience is achieved (or not achieved) by the characters in two of Turner Hospital's works, *The Last Magician* (1992) and *Oyster* (1996). To this end, I intend to explore in some depth the role played by narrative, and how it can either become an asset that allows resilience to develop, or else turn into a destructive force that leads to trauma. In the second place, I will explore more briefly the role played by love and end up by offering a reflection on ecological resilience. But, since both texts are rather long and complex, I will provide a brief summary of each to make things a little easier for the reader. *The Last Magician* revolves around a group of friends, Cat, Charlie and Catherine, who, as children, witness the only half-accidental death of retarded Willie, Cat's little brother. This death is in fact at least partially, if unwillingly, caused by an older child, Robbie, in his desire to impress Cat. As a result of these events, conveniently manipulated by Robbie and his rich and influential father, Cat begins her never-ending pilgrimage which will lead her to reform school in the first place and, as an adult who earns her living as a prostitute, from one brothel to another—and even to prison. She, however, always leaves a trail behind so that her friends can trace her until, abruptly, they lose track of her. The plot revolves basically around this search, in which two other people, Robbie's son Gabriel and his girlfriend, Lucy the narrator, also take part.

Oyster tells the story of how the outback town of Outer Maroo, which has always "managed to keep itself off maps" (4) in order to protect its inhabitants' uncomfortable secrets as well as their economy, is invaded, with the occasion of the turn of the century, by the members of a cult whose leader is the enigmatic Oyster. After a time of apparent normality, people begin to die and disappear both in the town and in the spot called the Reef, where Oyster has set up his camp, conveniently near the opal mines where he forces his followers to work for his own benefit. Other powerful people in town also have a share in the profits derived from this illegal business. The rest keep silent even after scores of people, including apparently Oyster himself, have died due to an explosion the origin of which nobody knows with any certainty. But then Sarah and Nick arrive in town, the former looking for her step-daughter, the latter for his son, both of whom have died in the catastrophe. Aided by Jess the narrator and by Major Miner, as well as by young Mercy Given, they will finally find out what has actually happened. Although nobody is able to prevent a second fire that destroys the town, along with most of its inhabitants, Mercy manages to escape with Nick and Sarah and drive towards a hopefully better future.

To return to the question of narrative, the first thing that has to be noticed is that many of Turner Hospital's characters are, in different ways, narrators. This is perhaps particularly remarkable in *The Last Magician* where, apart from Lucy, who writes the story we are reading, we encounter Charlie, the Chinese-Australian photographer, Gabriel, always intent on his note-taking, by means of which he hopes to reconstruct the truth, and

Catherine, who makes TV documentaries. This over-proliferation of narratives can perhaps be accounted for as attempts by the characters to carry out their mourning processes. Both *The Last Magician* and *Oyster* start with an abundance of unfinished business—so unfinished that the protagonists don't want to acknowledge that loved ones have died—friends, partners, daughters and sons: those who have read the novels will no doubt recall the impressive list which includes the cases of Cat, Charlie and Gabriel in *The Last Magician* as well as those of Brian, Amy, Angelo, Susannah, the Bugger and the whole of Oyster's community, Oyster included, in the homonymous novel. These are, Russell West points out, "the traumatic events out of which the narratives flow, and without which narrative would not be initiated." These resulting narratives, however, lack, he seems to believe, any healing power, because "the traumatic force [of the events] is such that they can never be related with any semblance of factual objectivity. Such events can be told from several mutually exclusive narrative vantage-points, and often at several removes" (2001, 184). While this is, to some extent, true, it must also be noted that, when the novels come to an end, a lot of the (not too many) survivors have managed, in different degrees, to come to terms with the truth and are ready to start out on fresh paths. This seems to suggest that if narrations are so convoluted, especially in terms of temporal structure, it may be due to the fact that the process of narration itself should be regarded, not only, or not so much, as being the product of trauma, but rather as a form of mourning whereby the subject is able to specify and gain some distance from her/his traumatic loss (LaCapra 2001, 66). No narratives are possible at the beginning of either novel because there are painful truths that will not be revealed. At the end, however, after covering the same ground over and over again in ways which only the attentive reader can follow, the result is that the actual level of knowledge the characters have now is, basically, not so different from the one they had at the beginning. However, as we are told on several occasions near the ending of *The Last Magician*, "Everything has changed and nothing has changed" (Turner Hospital 1992, 341). Narrative itself seems therefore to be both revealing and healing. Yet, the only thing that has actually changed is that we have accompanied the characters in the process of discovery: we have given them time; that necessary condition for wounds to heal.[2]

This need to cover the same ground once and again is epitomised by Charlie's remark that he takes pictures "so I'll see what I've seen" (Turner Hospital 1992, 36), a statement that sounds as if it was intended to explain the very concept of retrodetermination. Not being able to go through the "non-experience" of trauma, as Caruth would phrase it (1995, 7) he needs to capture it in retrospect. A characteristic example of this need to "see twice" is offered in the following passage:

> Charlie's body hummed with excitement. He was looking at another
> photograph of the bar in The Shaky Landing and the woman in the

corner was there again. [. . .] He couldn't understand why he hadn't seen her when they were there, but she must have seen him first and slipped away, she must have hidden for reasons of her own. It was the only explanation, he believed. (Turner Hospital 1992, 283)

Lucy is, however, a little more honest, when offering a different, though hesitant, explanation: "But the truth is, we don't know anything, and even less do we want to know. We don't know anything" (Turner Hospital 1992, 347).

Here is the well-known tension, so characteristic of trauma, between the need to know and the impulse to deny. And Charlie comes even closer to the truth when he says that "we know the answers to the burning questions but we are afraid of them, and so we need a screen. We need to project explanations and read them back" (Turner Hospital 1992, 14). Because, after all, as Kate Temby has observed, in *The Last Magician* "it is not an absence of answers that engenders the arduous journey to the centre of the labyrinth, but the impossibility of accepting the answers which on one level have always been known, yet which are constantly denied" (52).

Narrative, then, serves not so much to uncover the truth as to help characters to accept that they do know something, something that they have in fact known all along but which was simply too painful to accept in too straightforward a way. Now, what is actually at stake in this decision between accepting or refusing to know the truth is whether trauma or resilience will gain the upper hand.[3] Charlie, for example, decides to continue to ignore the truth even though he is surrounded by very powerful suggestions, not only that her childhood friend Cat is dead but, in addition, that she was cruelly murdered by yet another childhood friend. Thus, although he has of course received the impact of the first phase of trauma, that of aggression, of the very real loss of his friend, he refuses to accept the second blow, the one delivered by narrative and which would, from a Freudian point of view, unleash trauma. He prefers to stay, as Lucy once phrases it, "on the far side of the gaping space between seeing and interpreting" (Turner Hospital 1992, 344).

I believe, however, that it is precisely this refusal to suffer the blow of narrative that actually provokes the deeply traumatic state he lives in. He pretends to pursue the truth but chooses to interpret even the most obvious events in ways which amount to a denial of the truth, of that "answer to his burning questions" that he knows, in his heart of hearts, to be the only possible one. From this point of view, trauma means being suspended in the middle of nowhere, it means avoiding. This is not surprising in a novel in which, from the very first page, the narrator presents knowledge as a threat: "this knowledge engulfed me, a thick sack over the head" (Turner Hospital 1992, 3). This supreme paradox which presents knowledge as darkness prevents any progress towards the healing of trauma.[4]

If only avoidance follows the first blow, nothing can be done. In the case of Charlie, at least, it is, in fact, the process of resilience, rather than that of trauma, that requires that this second blow be delivered. By deciding not to know (a decision that he is ready to maintain even if he has to pay the price of his own life), he has also rejected the possibility of resilience, which could have broken the cycle of mere repetition, the habit of being installed in suffering. I am aware that this may sound daring, even cruel. I may perhaps clarify my meaning by introducing the distinction between the concepts of pain and suffering which, to the best of my knowledge, was first drawn up by Anthony de Mello:

> Pain exists, and suffering only arises when you try to oppose your pain. If you accept your pain, suffering does not exist. Pain is not unbearable, because it has an understandable meaning where it can find repose. What is unbearable is to have your body here and your mind in the past or in the future. What is unbearable is to try to distort reality. (1987, 29; my translation)

Living through one's pain is thus the only safe path towards true resilience. Most of the characters in the novel, however, want, as they often put it themselves, amnesia. Catherine, in particular, seems to have adopted the "I-want-amnesia" obsession as her motto, and her case is particularly curious in the sense that, as shown at the beginning of this chapter she devotes herself professionally to investigating and recording other people's traumas. She thus lives a vicarious life, surrounded by the pain of others, delving into it, so as to avoid looking her own pain in the face. Put another way, what looks like a desire to accompany the victims' painful processes is just a mask to cover over her hunger for amnesia, that highest degree of avoidance. In tune with this behaviour, she also intervenes as little as she can in her own programmes, an attitude with which she probably manages to take in everybody except, perhaps, herself.[5] What she is after by listening to so many unfortunate people is, at bottom, to not hear herself, to deafen herself to her own internal voices, concealing her pain from herself. Contact with others prevents her from having real contact with herself and with the true narrative of her own life.

Yet, contradictory as it may sound, what both Catherine and other characters procure by this procedure is to continue suffering rather than go through the experience of pain and have done with it. Of course, this choice is not without advantages. As somewhat bluntly put by Guillermo Borja:

> Suffering is a diseased content [. . .] It is a masochistic clinging to living badly, to repeating, because one is addicted to discomfort, both internal and external. Suffering avoids contact with pain, we suffer rather than accept and feel pain. Suffering is an external layer [. . .] Pain is contact with what we feel, with our lacks, with our essence.

[. . .] Suffering is noisy, while pain is silent, quiet. It is a state of solitude. Suffering is exhibitionist, it wants witnesses in front of whom to perform one's heroic acts. [. . .] Suffering is euphoric. (1995, 54; my translation)

If we accept this, it is not difficult to understand why, as Lucy says, "people run away from pain, I know that. I bolted from Brisbane once. Catherine fled to London. Charlie to New York"—where he, by the way, stayed for twenty-five years, no less, in the vain hope of appeasing his memories (Turner Hospital 1992, 311). It is arresting, to say the least, to find Charlie in his desperate flight from pain juxtaposed with the (theoretically at least) object of his search, Cat, who, by "[sculpting] her body into an artefact of abuse [. . .] makes a monument of her own pain" (301). Afraid of pain as he is, Charlie remains caught in the nets of repetition. He doesn't really want to find out the truth, even though, as Lucy informs us, "he himself took photographs obsessively in order to see what he had seen" (229). He is not successful for the simple reason that, he does not, albeit at a quite unconscious level, want the truth. Living with that knowledge would be far too painful for him to bear. It has to be conceded that he would be happy to provide the key to solving Cat's murder,[6] and from that point of view, he is morally redeemed: redeemed from a crime he never committed but of which he feels guilty. Otherwise, all he manages in his alleged pursuit of the right photograph is to get himself killed by frequenting the dangerous night joints he and Gabriel have repeatedly been warned off visiting. I do not think it is farfetched to argue that he unconsciously chooses death rather than a truth that it is too painful for him to come to terms with.

Ironically enough, he has, on another occasion, actually ensnared a criminal by means of the narrative told by his pictures, when a paedophile, watching one of Charlie's exhibitions, believes himself to be found out, and so ends up committing suicide right before the police arrest him. That experience encourages him to believe, later, "that all he had to do was wait, though the taking of photographs, he knew, was germane to the plot" (Turner Hospital 1992, 238). He therefore concludes that

the moral of the story [. . .] is patience. You have to *wait*.
Silence, exile, and cunning, he said.
And patience.
And photographs. (242)

But when it comes to clearing up who killed Cat, all of his picture-taking efforts are to no avail. In an unexpected reversal, it is a photograph of Sheba that Robbie, the murderer himself, took, which provides the final evidence against him. The picture seems to be done in Charlie's style, to such an extent that it can deceive even a connoisseur like Lucy, who takes it for granted that "it's unmistakably stamped with Charlie's mark" (Turner

Hospital 1992, 237). But the eye behind the camera was in fact Robbie's, thus proving right Sheba's contention that "[she is] a mirror" (338) and that, therefore, "we're not looking at *me*, you drip. We're looking at the blokes who took the pics" (326), "when you look at me, you see the man who holds the camera" (338). And, indeed, when Robbie looks at the picture, what he sees is not Sheba, but his own image as a murderer instead: "So you know" (339, 340), he says to Lucy the very moment he sees the picture, although Lucy is, at that point, unable (or, as I prefer to put it, unconsciously unwilling) to understand that his words entail a confession. In a sense, Robbie and Charlie can then be said to share, apart from their obsession with Cat, an unconscious drive towards failure.[7] Robbie, on the one hand, has very likely been taking and sharing the pictures that betray his obsession (and thus, his possible implication in the crime) in order to be discovered, and therefore relieved from the weight of guilt. Charlie, on the other, looks for death as the only way out that will allow him to keep up his fake narrative of hope until the end. He thus chooses the most radical way of numbing himself for good.

There are, however, other possible choices, as illustrated by Lucy. Although she, too, has managed not to feel anything for a long time, she will eventually be able to break the cycle of repetition towards resilience by deciding to live through her pain, perhaps because she feels that there are "matters far more disturbing than grief" (Turner Hospital 1992, 320). She is, though, far too used to avoiding pain, even before Gabriel's disappearance takes place: "I had meant bail out of attachments, bail out of coming pain. I flew solo and I always had. I was unhurtable. I stayed clear of everyone's nets" (282–83). To make things worse, her friends are not ready to keep her company in the process of working through her pain. It is no wonder that she encounters such difficulties given the social weight of pain, grief, mourning, as described by Cyrulnik: "When the narratives of those who surround the wounded person make them keep quiet and insist that nothing has happened, they remain numbed and get on by suppressing the chapter of their own history that other people cannot bear" (2009, 70) This is exactly what happens to Lucy when, very much in need of talking about her pain, she realises that she cannot resort to the help of her otherwise loyal friend Sheba: "I also knew instinctively that I wasn't equal to one of Sheba's 'pull yourself together' talks'. I wanted to brood disgracefully. I wanted to sit in my hotel room and feel Charlie's photographs in my hands" (Turner Hospital 1992, 337). Thus, she experiences how, to borrow Cyrulnik's words again, "a narrative that cannot be shared tears relationships up," since "what was possible in the sphere of the real becomes impossible to represent. That intimacy cannot be shared [. . .]. Such a narrative expels one from the community of representations that can be shared and that constitute a particular culture" (2009, 216, 204). She comes very close to trying to suppress her own narrative when, instead of legitimising her feelings and her history, she seems to take sides with those who have no patience to listen to her: "I know there's nothing more tedious than someone else's grief or state of shock" (Turner Hospital 1992, 333). Similarly, in this example, she also tends to internalise other people's point of view when labelling her need for a proper mourning process as just "brooding disgracefully."

She will, however, eventually prove able to break the cycle of repetition towards resilience and get moving after having lived through the paralysis of suffering, by virtue of the redemptive powers of pain. Lucy's narrative is different from Charlie's in that it leads her to acceptance. After having gone through all the circles in her Dantean hell, she finally comes to accept the truth that her dead (her boyfriend as well as her best friend, and the mythical figure of Cat, who is her double in more than one way) are dead and that they were killed by a man she had come to regard as her friend. Bitter indeed as knowledge (or rather, the acceptance of the knowledge she already possessed from the very beginning of the novel) turns out to be, it constitutes the only way in which she can finally shape the key narrative of her life and therefore provide her existence not only with a structure that can support the whole weight of her past, but also open up altogether new paths for her future. By living out her pain through the convoluted 350-odd page narration she has just concluded, she has actually managed, not only to express her pain, but also to actually live through that pain, and therefore to free herself. Her life will be thoroughly transformed from now on: her search for her lost friends has finally been brought to an end, and she is ready to devote her energy to more fruitful purposes.

In the first place, she decides to go back home to Australia, to the country she has been avoiding for years, and specifically to the rainforest, the site that has been haunting her imagination. She is now able to understand that "I've never really lived anywhere but Queensland and it's time to come home" (Turner Hospital 1992, 352). Her homecoming is particularly significant after her previous frantic life, in perpetual movement. In the first few pages, in fact, the reader is likely to feel some degree of vertigo, since we encounter a narrator who, speaking from a Sydney in which she has just arrived from London after some months spent in New York and Boston, calls her friend Catherine back in London. As a matter of fact, she changes places so frequently that at one point she will be unable to decide whether a particular event took place in London or in Sydney. "Both, it seems to me now," she concludes (57). What is at stake here, as I hope will have by now become obvious, is not just literal homecoming, but also a return to herself after her traumatic experience. After spending the whole novel running, or flying, from place to place, she has finally gathered enough courage to abandon her avoidance strategies and keep quiet in order to have the blow of knowledge delivered to her, thus leaving it open to the possibility of healing. This proves Cyrulnik's point that the narrative of one's own history is not necessarily a return of the past, but can also become a reconciliation with one's own history.

In the second place, freed from her own ghosts, she can now start think-ing of doing something useful for other people, the most healing of expe-riences for a victim of trauma according to Cyrulnik (2001, 214), and so she decides to make two new documentaries.[8] One of them will deal with Aboriginal land rights and the other with the inhabitants of the quarry, the novel's imaginary space for the very real places where people die of starva-tion, drug addiction, or another of the many plagues that accompany pov-erty and marginalisation. Lucy has, therefore, been able to overcome her personal trauma in order to help alleviate the collective trauma of the dis-possessed peoples of Australia. It may be helpful to remember here the dif-ferent attitudes that, Cyrulnik points out, may be taken up after a trauma has occurred: "One can abandon oneself to suffering, try to be indifferent or devote oneself to being a victim": all of them attitudes that, at different points in *The Last Magician*, are abundantly displayed by Charlie, Cath-erine and Lucy, and even Robbie. They are all, however, "anti-resilient, because they all mean an obstacle for any process of development" (2009, 128). As opposed to this, one can "do something with that suffering, use the need to understand it in order to transcend it and transform it into a social or cultural project," an attitude that "promotes resilience" (2009, 128–29), and which is precisely the one Lucy chooses as the novel comes to a close.[9]

It can therefore be concluded that narratives of suffering, like Char-lie's, are sterile and can only perpetuate trauma and potentially even lead to death, while narratives of pain, like Lucy's, are fruitful, transforming and life-bearing. As beautifully put by the Spanish writer Antonio Gala: "I won't take French leave of pain and get out through the back door. I'll inhabit it and allow myself to be inhabited by it. There is no other way to make our house bigger so that, when joy comes, if it does come, it will have more space in which to blossom" (1997, 20–21; my translation).[10]

And what, one might ask here, is the difference between Charlie and Lucy, which makes resilience possible for the latter, but not for the former? As was shown previously, Cyrulnik considers that the creation of loving bonds, especially when one is a child, are essential in order to give resilience a chance. The importance that he attaches to love is expressed eloquently enough by the Spanish translation of the title of one of his books, *El amor que nos cura* (translated literally as "The Love that Heals us" and pub-lished in English with the title *Talking of Love*) which, to my mind, per-fectly expresses the spirit of the author's message. Such is the importance of love as a factor of resilience that, Cyrulnik insists, the encounter with one single significant person is enough to make it possible for a person to get on in life, since "whatever allows a resumption of the social link will allow the wounded person to reorganise the image s/he has of herself/himself" (2001, 214), and this is true even if the relationship is very brief, because it may last a long time in the wounded person's memory, "and it is precisely there that identity is constructed" (168–69).

One might argue that this contradicts what actually happens in the novels, since Lucia was abandoned by her parents as a baby, while Charlie has been brought up by loving parents who try to provide him with everything he needs. Yet, Charlie grows up with different, contradictory narratives about himself, which affect the very core of his identity. To his parents, he is not only an object of love, but also "a true-blue Aussie," since he was born in Innisfail; to his schoolmates, on the contrary, he is "a yellow wog from China" who deserves nothing more than to be despised and beaten up. In addition to this, he grows up trapped between the mystic-heroic narratives of traditional Chinese culture as transmitted by his parents and the myth of "our [an "our" that, of course, excludes Charlie, who is consistently treated as a foreigner] Australian commitment to fair play" (Turner Hospital 1992, 217), as put by the "kindly fatherly judge" (218) who decides that Cat is to go to reform school. To phrase it in Cyrulnik's terms, Charlie is a victim of, "the narratives [he has] grown up with, those of [his] family and those of [his] culture," and which have "[instilled] into his soul a representation of himself disturbed by social myths" (2009, 21).

Charlie's parents are Chinese immigrants whose obsession is to enable their bright only child to achieve a university education, and perhaps they devote too much attention to work and do not have time for the little gestures that strengthen real personal links. Thus, when Cat first kisses Charlie, "he feels as though he has been pitched over the cliff in Wang Wei's painting and is soaring through sky. He has never keen kissed on the lips. His father, very occasionally, touches him on the shoulder. When he goes to bed, his mother puts both hands on his shoulders and presses her lips lightly and briefly against his forehead" (Turner Hospital 1992, 171–72). Lucy, as a child, is, in pointed contrast, "everyone's Little Wonder, the emperor's nightingale" (37). As she ironically describes herself, "she was still Lucia Barclay then, immaculate in the uniform of one of Brisbane's best private high schools for girls, a senior, a prefect, a winner of academic trophies, sports trophies, debating club trophies, [. . .] the flower of her school where she discussed Virgil with the Latin mistress, [. . .] where she had elegant Sunday dinners with the headmistress" (38). Thus, the joint forces of narrative and love have favoured Lucy's ability to develop resilience but Charlie, as the inheritor or a very different sort of narrative, as well as of a lack of loving physical contact and kind words, falls a prey to trauma. This explains why Charlie chooses, as has been explained, death, while Lucy can conclude that "hope and love are all we have and they are very potent baggage for people who travel light" (350).

Constance, too, seems sometimes to be presented as an example of resilience through love. After the tragic ending of her marriage to Robbie, she finds a new partner with whom she now lives, as was her desire, growing pineapples in the rainforest. There, she leads a life which looks like an example of mature serenity, and, in fact, she has been able to forgive the suffering Robbie inflicted on her: "You should never remember someone in rage or panic [. . .]. It isn't fair to them, they're not themselves. The way I remember Robbie is sitting on the end of Gabriel's bed telling him stories" (Turner Hospital 1992, 346). Again, the way she has chosen to construct the narrative of her life with the help of potent affectional bonds seems to have opened up the way to resilience. The way she has managed it is one often sanctioned by Cyrulnik: by eliminating some of the pieces of the total picture and choosing instead to highlight others, because, "when the real is monstrous, you have to transform it so that it will be bearable" (2001, 156). Constance very openly confesses to this manipulation of the truth with a view to surviving: "I used to feel I had to know what happened [. . .] but then gradually, it didn't matter anymore" (Turner Hospital 1992, 344). Constance's happiness begins to sound questionable when she starts to sound just like Catherine: "After a time it didn't matter. You just stop gnawing away at puzzles that don't make sense. You want amnesia" (345–46). And, finally, the inconsistency of her resilience mechanisms is clearly shown when, after telling the story of how a body (Cat's, in all probability) was found at the falls, she concludes that "it's your own blackness. It breeds black ideas. [. . .] Lucy, [. . .] I had to tell someone, it's been strangling me" (347). No matter how much she tries to suffocate her real feelings, the contradictions between what she would like reality to be and what she knows it to be like are but too patent here. Rather than real resilience, therefore, Constance illustrates denial, but, as Cyrulnik explains, "turning away from thought so as to prevent traumatic images from getting fixed in one's mind is a defence mechanism that does not set resilience at work" (2009, 60).

The previous examples are, therefore, more make-believe than real resilience. They involve a merely mental process, a new way of avoiding the "burning questions." That sort of well-being, if it deserves that name, can only be achieved by growing apart from reality. Perhaps a better example of real resilience is that offered by Jess and Major Miner in *Oyster*. They have the courage to start a relationship even if they are surrounded by the bushfire and there are only three survivors, even if, as Jess phrases it, "the end of the world is upon [them]" (Turner Hospital 1996, 43) and it all looks as if it is "the Day of Wrath" (45). The state of affairs could not be worse. But then, right at the very core of destruction, Major Miner states: "On the plus side [. . .], I've survived a few Armageddons and kept going. You could say I take ends-of-the-world in my stride. Almost, anyway. I suppose that's something" (50). That his words are more than just mere words is shown by the beautiful love scene that follows immediately after, and which Jess comments on in the following terms:

Beginnings astonish me, the way they can rise out of ashes; and as for the histories of lovers, they're outrageous [. . .] not to mention the question of their ruthlessness, of their swimming through joy like heedless kids while the end of the world is taking place, fiddling each other while Rome burns.

The sheer tactlessness of starting over at such a time!

This is the sober truth: *In the beginning* is always now, and ever shall be, world without end, amen. (50)

Once again, love and the construction of stories create the right atmosphere for resilience; sex and story-telling become one reminding us once again that, in Cyrulnik's words, "working on meaning is the most private of activities" (2005, 18). And I find this especially moving in a story whose protagonists are two old people who had decided to cut off human contact forever. Yet we now find them "training for new relational skills, working on the history that constitutes [their] identity, learning to think of [themselves] in different ways and fighting the stereotypes that our culture trots out about the wounded," all of which amounts, Cyrulnik says, to "the ethical commitment of resilience" (2005, 121).

But, at this troubled beginning of the twenty-first century, we can hardly talk about human resilience without making at least a brief but necessary reference to ecological resilience, since the former is impossible without the latter. Turner Hospital has also made room for such a concern in *Oyster*, especially by means of the presence of the Murris, the Aboriginal peoples who inhabit the Queensland area, and who are posed as an example of extraordinary resilience. As Ethel, the only Aboriginal character whose voice we are allowed to hear in the novel, puts it: "'Fuck off, Jess', she grins. 'Whitefella Maroo been and gone once, and been and gone twice, and we're still here, my mob and me'" (Turner Hospital 1996, 44). Such resilience (perfectly illustrated by Ethel's attitude when "putting the scattered rocks" of the bora rings "back where they belong, filling gaps in the circles and centuries," [50]) is, the text seems to imply, due to the Aboriginal belief that, in Deborah Bird Rose's words, "the lives of all Australian people are inextricably bound together, as are the soils, water systems, and the lives of plants and animals" (24). The incredible resilience of Aboriginal peoples is therefore due to the attitude that David Gilcrest would call "an ecocentric ethic of interconnectedness" (quoted in Bryson 2005, 2).

Even if, however, human beings give up their responsibility, there is something inherently resilient in nature, as Nick resolutely states in a few words, which, I find, constitute a complete treatise not only on human, but also on ecological resilience: "You can't pollute the ocean [. . .]. It just throws everything back out on the shore eventually, even oil slicks. That comforts me. That there's something, you know, that goes on resisting" (Turner Hospital 1996, 439). This comment is still echoing in the air at the moment when Sarah finally gives way to pain, overcoming the temptation of suffering and thus opens up the possibility of a successful mourning process for her dead stepdaughter. At this point, human and ecological resilience are, it seems to me, beautifully brought together:

> Her sobbing is violent but noiseless. [Nick] holds her. 'Listen', he says. 'The ocean . . . Think of the ocean.'
>
> Her sobbing is noisy now. It bounces off the rock walls and reverberates and echoes back from deeper down. An ocean of mourning fills the tunnel. They sit in the small cleaned sand-coloured space and listen to the dirge of it. The light from the torch washes them. He strokes her hair. He kisses her. They huddle like frightened children, holding each other, and stare into the dark. (443)

It is not by chance, then, that the ending of the novel, however inconclusive, teems with images of the ocean. In fact, it comes to an end with Mercy, the sixteen-year-old protagonist, driving precisely towards the coast, a world very different from the only one she has known so far: the Australian outback. Instead, she is now moving towards "Brisbane, the golden city. She imagines the great river with water in it. She thinks of grass, ferns, trees, ocean, sand. She imagines herself running into the ocean as into the world. She will let the world crest and froth about her" (453). Both on the ecological and the individual levels there is something, and somebody, that, as Nick would have it, goes on resisting. Cyrulnik, however, would probably not approve of the word "resisting" here, since he introduces a very useful distinction between resistance and resilience. Resistance is, according to him, often connected with defensive but hostile counter-aggression, which amounts in fact to a legitimisation of violence and opens up the way to mutual destruction. This phenomenon, he argues, "can be called 'resistance', but is certainly not 'resilience', since the past, by repeating itself, prevents a new line of development" (2009, 76). In other words, it could be argued that resilience is inwardly oriented: it is oriented exclusively towards yourself, it has got to do with the way you want to live, with how you cope with your own affairs[11] and whether you actually want to work for yourself, while resistance has to do with keeping an outward orientation, bent on aggression towards the enemy, on revenge, on placing your energy outside rather than inside.

Let us then call Mercy's attitude resilience rather than resistance, since it implies growth and transformation. It is highly significant that such resilient processes take place in an epilogue which Turner Hospital chose to entitle nothing less that "The End of the World." Those readers familiar with her works cannot forget her love of Eastern culture, and in particular, of the Chinese philosopher Lao-Tzu, whose wisdom often reverberates through her novels. At this point, I cannot help recalling his words, which I would like to use to end my argument with: "What the caterpillar calls the end of the world, everybody else calls a butterfly."

*The research carried out for the writing of this chapter is part of a project financed by the Spanish Ministry of Economy and Competitiveness (MINECO) (code FFI2012–32719). The author is also grateful for the support of the Government of Aragón and the European Social Fund (ESF) (code H05).

NOTES

1. Quotations from Cyrulnik's works are my translation, except in the case of *Talking of Love*, where the English translation has been used.
2. Catherine is also very much aware of that need. When Lucy was earning her life as a prostitute, one of her customers, who is turned on by watching Catherine's programmes, remarked that "you see, she won't cut in, she lets [her interviewee] take his own time, she'll let him take all the time in the world if he needs . . ." (Turner Hospital 1996, 124) in order to construct the narrative of his traumatic experience.
3. The relationship between truth and resilience, it must be noted in passing, is not always as straightforward as this, since, as Cyrulnik highlights, there are occasions on which a memory which is "too real and remains uninterpreted may prevent the process of resilience from developing" (2001, 156).
4. The fact that the pull towards resilience finally gains the upper hand, at least as far as Lucy is concerned, is hinted by the choice of the very last word in the novel: "light" (Turner Hospital 1996, 352).
5. It should at least be acknowledged, in Catherine's favour, that eventually she will be able to talk about her childhood's traumatic events, which have marked the whole of her life. Her ability to communicate is, however, restricted to communicating with Lucy, to whom she says repeatedly that "we're like war vets" (e.g., Turner Hospital 1996, 321), just like the one we have seen Catherine interviewing, because they "can only talk to each other." Such a restriction is only natural, according to Cyrulnik, since "survivors feel that the only people who can really understand them are those who have also survived" (2009, 175). Catherine's formulation is also interesting in that it returns to the well-known early attempts at identifying the origins of trauma with post-war experiences such as shell-shock. It somehow seems to me that Catherine could not legitimise her own feelings, as if, from her point of view (one which contradicts her own experience), only those involved in big world-size events were entitled to traumatic pain. There are, however, abundant studies (see Ann Cvetkovich 2003, 16; Forter 2007, 260) that have already provided solid evidence that trauma may even be connected to events in personal life, which would be regarded as very ordinary-looking by many. Additionally, numerous authors have discussed the insidious nature of traumas woven into the very fabric of our lives, such as those related to gender or class issues—including Laura Brown, who first introduced the formulation "insidious trauma" (1995) as well as Laurie Vickroy (2002, 10) or Cvetkovich (2003, 17).
6. And even this can be questioned, since his feelings towards Robbie are ambivalent enough to constitute an obstacle to Charlie's apparent desire to find out who committed the crime.
7. This parallelism is disturbing, since it brings far too close the figures of victim and perpetrator, creating an ambivalence that is often to be found in trauma studies (see, Kalí Tal 1995, 10; Vickroy 2002, 19).
8. It must here be mentioned that these documentaries, which are to be made once Lucy has already come to terms with herself, have little to do with those mentioned earlier on in this chapter, and which are merely Catherine's attempt at avoidance.
9. While, at the beginning of the novel, she was already in the habit of spending time with the inhabitants of the quarry, she did so basically from a feeling of rage towards the injustice that such things could come to happen in the society she lives in, but she never took a positive step towards changing them.

Now, however, she has assumed a position of power, and her new documentaries promise to become exactly what Cyrulnik refers to: a complete social project, able to bring about change to both other people and herself.

10. The Spanish original reads: "No me despediré [del dolor] a la francesa y saldré por la puerta falsa. Lo habitaré y dejaré que él me habite. No hay otro modo de ensanchar nuestra casa y de que, cuando venga la alegría, si viene, tenga más sitio donde recrearse."

11. Technically, according to Cyrulnik, a difference must be established between "coping," which takes place simultaneously with the traumatic event, and "resilience," which develops after the traumatic occurrence (2009, 121).

WORKS CITED

Bird Rose, Deborah. 2000. *Dingo Makes Us Human: Life and Land in an Australian Aboriginal Culture*. Cambridge: Cambridge University Press.

Borja, Guillermo. 1995. *La locura lo cura: Manifiesto psicoterapéutico*. Vitoria: La Llave.

Bowlby, John. 1979. *Vínculos afectivos: Formación, desarrollo y pérdida*. Madrid: Morata.

Brown, Laura S. 1995. "Not outside the Range: One Feminist Perspective on Psychic Trauma." In *Trauma: Explorations in Memory*. Edited by Cathy Caruth,, 100–112. Baltimore and London: The John Hopkins University Press.

Bryson, J. Scott. 2005. *The West Side of Any Mountain: Place, Space and Ecopoetry*. Iowa City: University of Iowa Press.

Callahan, David. 2009. *Rainforest Narratives: The Work of Janette Turner Hospital*. St Lucia: University of Queensland Press.

Cvetkovich, Ann. 2003. *An Archive of Feelings: Trauma, Sexuality, and Lesbian Public Cultures*. Durham and London: Duke University Press.

Cyrulnik, Boris. 2001. *Los patitos feos. La resiliencia: una infancia infeliz no determina la vida*. Barcelona: Gedisa.

———. 2005. *Talking of Love: How to Overcome Trauma and Remake Your Life Story*. Translated by David Macey. London: Penguin Books.

———. 2009. *Autobiografía de un espantapájaros. Testimonios de resiliencia: el retorno a la vida*. Barcelona: Gedisa.

De Mello, Tony. 1987. "La iluminación es la espiritualidad." *Vida Nueva* 1583: 27–66.

Forter, Greg. 2007. "Freud, Faulkner, Caruth: Trauma and the Politics of Literary Form." *NARRATIVE* 15 (3): 259–85.

Gala, Antonio. 1997. *La regla de tres*. Barcelona: Planeta.

Greiner, Donald J. 2007. "Ideas of Order in Janette Turner Hospital's *Oyster*." *Critique-Studies in Contemporary Fiction* 48 (4): 381–90.

LaCapra, Dominick. 2001. *Writing History, Writing Trauma*. Baltimore and London: Johns Hopkins University Press.

Samuels, Selina, ed. 2006. *Australian Writers, 1975–2000* (Dictionary of Literary Biography 325). Detroit, Michigan: Gale.

Tal, Kalí. 1995. *Worlds of Hurt: Reading the Literatures of Trauma*. Cambridge: Cambridge University Press.

Temby, Kate. 1995. "Gender, Power and Postmodernism in *The Last Magician*." *Westerly: A Quarterly Review* 40 (3): 47–55.

Turner Hospital, Janette. 1988. *Charades*. St Lucia: University of Queensland Press.

———(published under the pseudonym Alex Juniper). (1990) 1994. *A Very Proper Death*. New York: Fawcett Crest.

———. 1992. *The Last Magician*. London: Virago.

———. 1996. *Oyster*. London: Virago.

Vickroy, Laurie. 2002. *Trauma and Survival in Contemporary Fiction*. Charlottesville and London: University of Virginia Press.

West, Russell. 2001. "Multiple Exposures": Spatial Dilemma of Postmodern Artistic Identity in the Fiction of Janette Turner Hospital." In *Flight from Certainty: The Dilemma of Identity and Exile*. Edited by Anne Luyat and Francine Tolron, 177–90. Amsterdam and New York: Rodopi.

8 The Burden of the Old Country's History on the Psyche of Dominican-American Migrants

Junot Díaz's *The Brief Wondrous Life of Oscar Wao*

Aitor Ibarrola-Armendáriz

Which is why it's important to remember *fukú* doesn't always strike like lightning. Sometimes it works patiently, drowning a nigger by degrees, like the Admiral or the U.S. in paddies outside of Saigon. Sometimes it's slow and sometimes it's fast. It's doom-ish in that way, makes it harder to put a finger on, to brace yourself against. But be assured: like Darkseid's Omega Effect, like Morgoth's bane, no matter how many turns and digressions this shit might take, it always—and I mean always—gets its man.

Junot Díaz, *The Brief Wondrous Life of Oscar Wao* (2007)[1]

The collective trauma works its way slowly and even insidiously into the awareness of those who suffer from it, so it does not have the quality of suddenness normally associated with "trauma." But it is a form of shock all the same, a gradual realisation that the community no longer exists as an effective source of support and that an important part of the self has disappeared.

Kai Erikson, *Everything in its Path* (1976)

INTRODUCTION: ON COLLECTIVE TRAUMA

Although from two radically different sources—one from a recent work of fiction by a Dominican-American writer and the other from a ground-breaking, socio-psychological study of the effects of natural disasters on survivors by an Austrian-born sociologist—the two epigraphs above refer to aspects that seem distinctive of collective traumas. On the one hand, Erikson's analysis of a major environmental catastrophe in the mountains of West Virginia underlines the slow and insidious way in which the effects of the cataclysmic event found their way into the minds of the victims, gradually disrupting the social networks and neighbourly practices of the community. On the other, Díaz's description of fukú (or "*Fukú americanus* [. . .] generally a curse or a doom of some kind; specifically the Curse and the Doom of the New World" [2007, 1]) puts the emphasis on the intangible, yet inevitable, manner in which these violent disturbances come to have a significant impact on successive generations.[2] Most scholars who have delved into the phenomenon of collective trauma would agree that the sluggishness and unawareness with which these catastrophes enter the consciousness of those afflicted are, indeed, among their most prominent features (Alexander 1987; Sztompka 1993). They argue that what begins as social distress caused by different kinds of disruptive incidents is gradually transformed into a cultural crisis that is experienced not only by the survivors but also by witnesses and onlookers receiving symbolic representations of those original events. Thus, what is called the "trauma process" can be seen to develop in the transition between the horror-inspiring events and the collective representations of them that the group will invariably generate as time goes by:

> Trauma is not the result of a group experiencing pain. It is the result of this acute discomfort entering into the core of the collectivity's sense of its own identity. Collective actors "decide" to represent social pain as a fundamental threat to their sense of who they are, where they came from, and where they want to go. (Alexander et al. 2004, 10)

Erikson's award-winning study of the socio-psychological aftereffects that the destruction of a coal waste reservoir and the subsequent flooding had on a community in the Appalachian Mountains has become a classic in the analysis of disaster response and collective cultural trauma. His ethnographic and sociological arguments demonstrated that the destructive event had left "indelible marks upon the group consciousness, marking their memories forever and changing their future identity in fundamental and irrevocable ways" (Alexander et al. 2004, 1). According to Erikson, what had most crucially characterised the people of this mountain community up to the moment of the "accident" had been the unstable balance between "their sense of independence on one hand, and a need for dependence on the other" (1976, 84–88). When this balance was upset by the overwhelming event, the whole community went into a rapid, downward spiral that affected not only their material and social well-being but, most importantly, their psychic and spiritual sense of place and comfort. Bearing in mind the extensive literature on collective trauma that has appeared since the mid-1970s, it could be argued that Erikson's analysis is probably flawed by a "naturalistic" conception of trauma; that is, a conception that claims that shocks or abrupt changes will produce a rational response—in this case, the lawsuit brought by the survivors—to try to recover or reconstruct their earlier condition. But, of course, we have learnt more recently that those memories and reactions to disasters or acts of extreme violence are not just responses to past events, "but are interpretative re-constructions that bear the imprint of local narrative conventions, cultural assumptions, discursive formations and practices, and social contexts of recall and commemoration" (Antze and Lambek 1996, vii; see, also, Jelin 2003). Thus, the "trauma process" of appeasement of and potential recovery from those distressing episodes will invariably be dependent on the adequacy of the narrative strategies chosen to make "claims" about particular social realities and the receptiveness—or empathy—that others show towards the demand for symbolic / affective or more literal reparation.

At first glance, Junot Díaz's much-acclaimed first novel, *The Brief Wondrous Life of Oscar Wao* (2007), hardly seems to fit into the narrative patterns that we would immediately associate with the conventional modes of representation of the trauma process.[3] To start with, although there are abundant references to historical figures and episodes throughout the novel, it is also evident that the author is using them creatively and never intends to employ them in a documentary or testimonial manner. As Díaz has explained in several interviews, he doesn't "like dealing with fact" and, although he also warns us of the dangers of cultural amnesia, he feels more comfortable employing these materials to investigate the "sloppy recesses of the human heart" (Adair-Hodges 2008). And to a great extent, this is what he does in his Pulitzer Prize-winning novel which, although focusing on a young, Dominican-American aspiring writer, could never be considered autobiographical in any literal sense. Oscar de León is an overweight, self-loathing nerd whose fundamental dreams are to become a science-fiction writer, "the Dominican J. R. R. Tolkien," and to find the love of his life. As a second generation Dominican-American growing up in Paterson, New Jersey, he soon discovers that life in his Latino barrio is not as full of promise as other members of the diaspora would want him to believe. In fact, Oscar realises that, given his physical appearance and his bizarre interests, he "couldn't have passed for Normal if he'd wanted to" in the context of his school and neighbourhood: "Dude wore his nerdiness like Jedi wore his light saber or Lensman her lens" (Díaz 2007, 21). The protagonist's obsession with science fiction, comics, cartoons and role-playing games not only turns him into an outcast at school, but also appears to condemn his repeated attempts at securing a girlfriend to failure—which is especially "traumatic" in the case of a Dominican, who is supposed to be invariably successful with women. Moreover, Oscar's sad story is further complicated by a suggestion made by the narrator, Yunior de las Casas, in the prologue to the book, concerning the notion that Oscar may be the victim of a family curse, or what he calls a "high-level fukú" (5), which, as noted in the first epigraph, will doom him to endless unhappiness. Although it is unclear at first how the violence and horrors experienced by his relatives in the old country could expand and have a tangible influence on the protagonist's life in New Jersey, it is evident that one of the author's intentions in the novel is to have the vexing history of the country constantly casting shadows on the "hero's" current misadventures. Thus, when the narrator is intrigued by the question of where Oscar's fondness for fantasy literature and science fiction comes from, he muses in a footnote:

> Where this outsized love of genre jumped off from no one quite seems to
> know. It might have been a consequence of being Antillean (who more

sci-fi than us?) or of living in the DR for the first couple of years of his life and then abruptly wrenchingly relocating to New Jersey—a single green card shifting not only worlds (from Third to First) but centuries (from almost no TV or electricity to plenty of both). After a transition like that I'm guessing only the most extreme scenarios could have satisfied. [. . .] Or was it something deeper, something ancestral? (21–22)

Be that as it may, what seems fairly undeniable is that Oscar's pathetic existence in the New Jersey of the 1980s and 90s will inevitably be tied to the tragic history of his family back in the Dominican Republic and the sense of uprootedness that derives from their having had to run away to a different country.[4] As Kakutani pointed out in an early review of the novel, beyond the tragicomic portrayal of a second-generation Dominican geek in the US, the book also offers "a harrowing meditation on public and private history and the burdens of familial history" (2007). In fact, the longest sections in Díaz's novel are not devoted to Oscar's difficult times in adapting and finding love in the "land of freedom and opportunity" but, rather, to his grandparents' and mother's tortuous and violence-ridden lives in the Dominican Republic during the "Trujillo Era" (1930–61) and after. No doubt, the main character's psychological make-up does not help much in terms of integrating into the community: "Oscar was a social introvert who trembled with fear during gym class and watched nerd British shows like *Doctor Who* and *Blake's 7*, [. . .] he used a lot of huge sounding nerd words like *indefatigable* and *ubiquitous* when talking to niggers who would barely graduate from high school" (22). Nevertheless, the author seems to be much more preoccupied with the question of how one's ancestral homeland—with its superstitions, politics, social inequalities, etc.—and family history come to reach insidiously, but also surely, into the lives of others who at one point or another had to abandon their native country. In the case of Oscar, it is clear that many of those shadows from the past fall upon him through his mother, Hypatía Belicia Cabral, mostly referred to as simply Beli, whose life story back in the DR was marked by the early loss of her parents and siblings, the brutal abuse received at the hands of a foster family and a difficult relationship with a gangster, which eventually force her to leave her people and homeland. But apart from this figure bridging the two worlds portrayed in Díaz's novel, the protagonist's unlucky love affairs and his engrossment in fantasy fiction and hardcore science fiction are plagued with subtle and overt references to the horrors that his progenitors experienced under Trujillo's dictatorship: "What more sci-fi than the Santo Domingo? What more fantasy than the Antilles?" (6). Indeed, what the reader manages to see through the small window of the terrors faced by a college student who has problems of weight, self-esteem and identity is the much larger picture of the sorrows befalling a whole generation, or two, of Dominicans who cannot easily free themselves from what the narrator calls a fukú (or "the Great American Doom"). As Yunior ironically explains early in the novel, even readers who are incredulous of these "ghostly presences" should be very cautious: "It is perfectly fine if you don't believe in these "superstitions." In fact, it's better than fine—it's perfect. Because no matter what you believe, fukú believes in you" (5).

Resistance Histories and Collective Trauma Narratives

According to Mansbach, one of the main achievements of Díaz's novel lies in his ability to straddle two extremely different worlds—the DR of his elders and the present-day United States—by moving backward and forward in time: "The shadow of Rafael Trujillo, the Dominican Republic's vicious dictator, looms large over the life of a family, as does fukú, an ancient curse that may or may not explain the misery, heartbreak, violence, and mother-child conflict that have characterized their last 60-plus years" (2007). In this sense, several scholars have pointed out that Díaz may easily be inscribed within a group of "hyphenated writers" from the Caribbean who have invented a new literary genre intended to give them a footing in the host culture by "re-membering" the incredible and terrifying experiences that their grandparents and parents underwent in the old country, as well as the migratory processes that usually followed.[5] Méndez has argued, for example, that these narratives are characterised by the return of traumatic memories and unresolved mourning that come to trouble the displaced communities, and indelibly mark their identity (2008, 173–74). As will be observed below, in *The Brief Wondrous Life of Oscar Wao* we find many of the features that have become commonplace in collective trauma narratives, such as fragmentation, displacement, hesitance, repetition and resistance. On examining the effects of traumatic impact on survivors and witnesses, Douglass and Vogler have remarked that, apart from exerting their power "long after its first impact," they tend to produce symptoms that are often "disguised or symbolic in their manifestations": "The effects frequently include what the therapists call "dissociation," or disorientation of the thinking process and panic at being possessed by what seems to be unthinkable, along with a loss of the ability to trust in any grounds for conventional reality" (2003, 10–11). Yunior's reconstruction of Oscar's story and his family's history reveals many of these symptoms, as he himself is afraid that the kind of curse that drove his friend to an attempted suicide first, and then his fatal end, may still be at work as he reflects on the immense burden all of them seem to carry:

> The Darkness. Some mornings he would wake up and not be able to get out of bed. Like he had a ten-ton weight on his chest. Like he was under acceleration forces. Would have been funny if it didn't hurt his heart so. Had dreams that he was wandering around the evil planet Gordo, searching for the parts for his crashed rocket, but all he encountered were burned-out ruins, each seething with new debilitating forms of radiation. (268)

Sacks has noted that, despite the uncommonly wise and charismatic writing, and the memorable finish of Díaz's novel, "it's also loose and unsure of itself and a great deal of its latent power escapes through the cracks of a creaky construction" (2008). He also claims that, due to its fractured and discontinuous structure, the novel feels as if it had been thrown together hastily—something quite surprising given that it took the author almost a decade to write it. Certainly, there will always be readers who, like this particular reviewer, feel a bit frustrated by the non-linear and jumpy rendition of the lives of three generations of the Cabral-de-León family. However, two important factors need to be taken into account in order to do justice to Yunior de las Casas' task of putting together his "hero's" and his progenitors' misadventures. On the one hand, he is aware from the start that he will not be able to depend on any official records of the events, since they are most likely to have been completely erased or highly biased in the interests of the perpetrators.[6] As Hanna correctly notes:

> Throughout the narration, Yunior self-consciously struggles and experiments with how best to accomplish his task because in the process of his research, as he attempts to uncover both the story of the family and the history of the nation, he is continually confronted with silences, gaps, and "páginas en blanco" left by the Trujillo regime. (2010, 498)

This scholar refers to Díaz's novel as a "resistance history" precisely because it has to rely on alternative sources in order to restore collective memories and undo the evident socio-political repression Dominicans were faced with over long decades. As Humphrey has remarked, "a key strategy to prevent the return of violence and stop the effects of past violence haunting individual and social relationships is to confront the past" (2002, 105). On the other hand, Lacan (1977), Caruth (1995) and others have demonstrated that traumatic memories and responses are governed by unconscious drives that will produce complex displacements and repetitions in the narratives that eventually emerge. According to these psychoanalytic theorists, the initial shock is only experienced irrationally and, consequently, cannot be fully grasped, although it will definitely come back to haunt and distress the victim. In Caruth's words, "if PTSD must be understood as a pathological symptom, then it is not so much a symptom of the unconscious, as it is a symptom of history. The traumatised, we might say, carry an impossible history within them, or they become themselves the symptom of a history that they cannot entirely possess" (1995, 5). Yunior fits quite squarely into this description of the typical trauma witness, for he is repeatedly overpowered by the surfeit of history and the outsize destinies he comes across in his research. As Díaz himself explained in an interview, Yunior seemed the right kind of narrator because the very story he is telling makes him grow uncertain: "He's conflicted and ambivalent about it" (Weich 2007). And then, after discussing at some length the multifarious kinds of lacunae, silences and absences in the novel, Díaz adds: "Part of Yunior's interest in the history is not only to fill in context and background, but also for him to understand why he's over here telling this story about two places" (Weich 2007). This might explain those many points in Yunior's narration where he takes time off again to question and even undercut his own authority, since he is not sure whether he is making the right choices or not:

> There are other beginnings certainly, better ones, to be sure—if you ask me I would have started when the Spaniards "discovered" the New World—or when the U.S. invaded Santo Domingo in 1916—but if this was the opening that the de Leóns chose for themselves, then who am I to question their historiography? (211)

Hanna remarks that these moments of uncertainty and hesitation in Díaz's narrator are critical because they alert "the reader to the fact that the story to follow will draw on quite different sources, creating a pastiche that attempts to capture the Caribbean diasporic experience" (2010, 500). Indeed, from the very first pages of the novel, we realise that the text incorporates elements of different popular and literary genres such as magical realism, punk-rock, superhero comics, classical tragedy and noir. As A. O. Scott sees it, the novel proves at times excessively uncontrolled and disorienting since, "within its relatively compact span, *The Brief Wondrous Life of Oscar Wao* contains an unruly multitude of styles and genres" (2007). Still, one can better see the sense of this multiplicity of perspectives and registers when one realises that the political and cultural violence of the past—which often takes the form of traumatic memories—is usually seen through the interpretative frames and discursive practices of the present time from which it is remembered (see Jelin 2003, 16). There will be parts of these acts of retrieval and preservation of the past that will remain mostly unintelligible to us, for not everybody will be familiar with those frames and practices, but they should still be included in the trauma process. As Díaz contends, "and yet we still manage to pull together a culture, a self, and a history. One of the reasons his [Yunior's] narration is intriguing to me as a writer is because what bedevils him as a narrator bedevils the entire project of what we would call the Caribbean. He's in good company" (Weich 2007).

Other Disorienting Features of Trauma Narratives

One other feature of Díaz's novel that readers and reviewers have found off-putting, or even bewildering, is the extensive use of footnotes that sometimes makes the book look more like an academic thesis or a scientific treatise than a work of fiction. Indeed, many of the notes in the novel are meant to provide the reader with information concerning some of the historical figures, events and locations referred to in the story, without which it would be difficult to understand certain incidents and behaviours:

For those of you who missed your mandatory two seconds of Dominican history: Trujillo, one of the twentieth century's most infamous
dictators, ruled the Dominican Republic between 1930 and 1961 with
an implacable ruthless brutality. A portly, sadistic, pig-eyed mulatto
who bleached his skin, wore platform shoes, and had a fondness for
Napoleon-era haberdashery, Trujillo (also known as El Jefe, the Failed
Cattle Thief, and Fuckface) came to control nearly every aspect of the
DR's political, cultural, social, and economic life through a potent (and
familiar) mixture of violence, intimidation, massacre, rape, co-opta-
tion, and terror; treated the country like it was a plantation and he was
the master. At first glance, he was just your prototypical Latin American caudillo, but his power was terminal in ways that few historians or
writers have ever truly captured or, I would argue, imagined. (2)

But, of course, although formally Yunior may employ some of the formulae
he is learning at Rutgers to document his research properly, it is clear from
the tone and the kind of evidence that the narrator provides in these asides
that he does not intend to endow these notes with the degree of unobjection-
able factuality that is characteristic of history-writing and historiography.
In his essay "Writing History, Writing Trauma," Dominick LaCapra talks
at some length about the importance of footnoting in any research paradigm that aspires to clarity and accuracy by using specific evidence to make
truth claims (2001, 5–7). As this author explains, "the note (footnote or
endnote) is the correlate of research, and its use as a referential component
of research is one criterion that serves to differentiate history from fiction"
(5–6). Although LaCapra admits later on that fiction *may* include referential
notes, they should never be deemed to have the same status as those used in
professional historiography, since, rather that providing evidence to make
truth claims, they tend to block reference "by taking one back into the text
with loop-like or laberynthine effects" (2001, 7). As a matter of fact, most
of the notes that Yunior inserts in his account tend to produce this type
of effect because they generally connect the lives of Oscar's relatives with
places, people, traditions, historical events, etc., with which we would never
have suspected there was any relation. So, for example, when Beli becomes
pregnant with the gangster's child, we are told that he is already married to
one of Trujillo's sisters (known as "La Fea"). Predictably, when this woman
learns that Beli is carrying her husband's child, she gets two huge policemen
to kidnap Oscar's mother and force Beli to have an abortion. After being
cruelly beaten up in a cane field and miscarrying, Beli is miraculously saved
by a golden-eyed mongoose that guides her out of the cane and tries to cheer
her up by informing her that she will have two other children in the future.
In a footnote on the next page, we read the following:

The Mongoose, one of the great unstable particles of the Universe and
also one of its greatest travelers. Accompanied humanity out of Africa

and after a long furlough in India jumped ship to the other India, a.k.a. the Caribbean. Since its earliest appearance in the written record—675 B.C.E., in a nameless scribe's letter to Ashurbanipal's father, Esarhaddon—the Mongoose has proven itself to be an enemy of kingly chariots, chains, and hierarchies. Believed to be an ally of Man. Many Watchers suspect that the Mongoose arrived to our world from another, but to date no evidence of such a migration has been unearthed. (151)

No one would think of this description of the small, predatory mammal as scientific, for in fact the narrator seems to show a preference for hearsay and folk knowledge over testable evidence. But, naturally, given the role that this creature plays in Belicia Cabral's story and, later on, in her son's, it may make sense to enshroud it in all the mythical lore included in the passage. As Antze and Lambek have argued, "there is a dialectical relationship between experience and narrative, between the narrating self and the narrated self. As humans, we draw on our experience to shape narratives about our lives, but equally, our identity and character are shaped by our narratives" (1996, xviii). In the case of *Oscar Wao*, it is evident that people emerge from and are the products of the stories about themselves—be they horror stories, feminist prose or sci-fi tales—, in as much as the stories derive from their lives. As Yunior insists again and again throughout the novel, one needs to develop this ability to put together those bits and pieces of history and memory into an (at least partly) coherent whole, if one wishes to render all those lives meaningful.[7] Right before Beli's uncanny meeting with the mongoose, the narrator confides to the reader:

Whether what follows was a figment of Beli's wracked imagination or something else altogether I cannot say. Even the Watcher has his silences, his páginas en blanco. Beyond the Source Wall few have ventured. But no matter what the truth, remember: Dominicans are Caribbean and therefore have an extraordinary tolerance for extreme phenomena. How else could we have survived what we have survived? (149)

Using Douglass and Vogler's terminology, Yunior de las Casas' position would respond quite fittingly to the category of "onlooker trauma" (2003, 10), that is, one who observes—thus, the suitability of his self-assigned role as a "Watcher"—the effects of traumatic events or one who is a descendant of victims. As these scholars explain, "the process of witness depends heavily on cultural values and meanings, and changes in cultural context have contributed significantly to changes in the discourse of witness in recent years" (Douglass and Vogler 2003, 10).

To further substantiate my hypothesis, let me look swiftly into two or three other aspects of Díaz's novel that have been criticised by reviewers, but which could be easily explained if one adopted the strategy of approaching the text as an instance of a "trauma narrative." Mansbach complains, for example, that "Yunior's identity is only revealed midway through the novel, well after the question of whose voice we are reading begins to grow distracting" (2007). This author attributes the questionable decision of keeping the narrator's identity veiled for so long to the fact that the writing process was excessively extended in time and "presumably full of cuts and rearrangements." However, perhaps the real intention behind this decision—not so uncommon in trauma narratives—is to turn the narrator into another member of the "carrier group"—that is, a representative or spokesman of the traumatised community who reveals "particular discursive talents for articulating their claims—for what might be called "meaning making"—in the public sphere" (see Alexander et al. 2004, 11). If this were the case, Yunior's personality traits, which are by the way radically different from Oscar's, and his connection with several of the main characters in the novel would not seem so central, since his major role would be as an active agent of the "trauma process" giving voice to the perspective of second-generation Dominican migrants in North America. Yunior's role in the story proper could be said to be marginal and, even, elusive, but his ability to construct a compelling framework of "cultural re-inscription" in which trauma can be finally represented is definitely essential. He seems to be perfectly equipped to engage in the work of providing those atrocious representations with new signification for the "carrier group," as well as to persuade the wider audience that they have also become "obliquely traumatized" by witnessing and experiencing the pain of the victims (Alexander et al. 2004, 12). Predictably, this task of meaning-making and empathy-arousing is by no means easy and it is not unusual to find the storyteller fairly exhausted when s/he has completed his/her new, alternative narrative: "It's almost done. Almost over. Only some final things to show you before your Watcher fulfills his cosmic duty and retires at last to the Blue Area of the Moon, not to be heard again until the Last Days" (329).

One other element in the novel, which may have been misinterpreted by critics, is the abundant references to beauty and sexuality, especially when the narrator talks about Dominican women. Beyond painting in words some captivating female portraits, the author seems intent on showing how, in the midst of the incredible brutality and sexism prevailing in the DR, there were still some rebellious women who managed to preserve their grace and independence—intellectual and otherwise. To provide a succinct example, here is Beli's introduction early in the novel:

> Before there was an American Story, before Paterson spread before Oscar and Lola like a dream, or the trumpets from the Island of our eviction had even sounded, there was their mother, Hypatía Belicia Cabral:
>   a girl so tall your leg bones ached just looking at her
> so dark it was as if the Creatrix had, in her making, blinked

> who, like her yet-to-be-born daughter, would come to exhibit a particular Jersey malaise—the inextinguishable longing for else-wheres. (77)

Although it may be true that the narrator occasionally indulges in a hip-hop machismo and some profanely oversexed language, the fact is that, as Scott underlines (2007), the novel centres very much on the unruly and courageous women in the Cabral-de-León family. Oscar's grandmother, Lydia, his mother, Belicia Cabral, and, of course, Beli's foster mother, La Inca, all have magnetic and highly resilient personalities that struggle to survive and blossom under the vicious rule of Rafael Trujillo. When Beli realises that she will need to abandon the island as a result of her affair with the gangster, and that the dictatorship is finally coming to an end, we begin to understand the kind of woman that her aunt-turned-into-mother is: "La Inca, who I don't think slept a single day during those months. La Inca, who carried a machete with her everywhere. Homegirl was 'bout about it. Knew that when Gondolin falls you don't wait around for the balrogs to tap on your door" (161).[8]

As a matter of fact, one could even speak of Oscar's thwarted romances and his stalling literary projects as being clearly foreshadowed by earlier family episodes that, in a way, "reverse the migratory path from the D.R. to the U.S.A." (Scott 2007), and end up with the protagonist meeting his fate in the land of his ancestors, a victim of love and of his anxiety to find himself. Somehow, Oscar's going back to his mother's land in search of his beloved Ybón Pimentel—an older prostitute—closes the circle of initial shocks, repression, repetition and return that is characteristic of trauma processes (see Caruth 1996, 106–7). His final days on the island are almost an exact replica of his mother's violent, heart-breaking and humiliating experiences just before she had to leave the country: "It says a lot about Beli that for *forty years* she never leaked word one about that period of her life: not to her mother, not to her friends, not to her lovers, not to the Gangster, not to her husband. And certainly not to her beloved children, Lola and Oscar. *Forty years*" (258). Belicia Cabral's attitude to her past—one of wilful denial and hiding—presents the perfect case study for trauma scholars, since she shows all the symptoms of those who try to put behind them, mostly without success, those wounds and disorienting losses that have indelibly marked their existence and identity. As becomes apparent in the closing pages of the novel, for Oscar too, there was no getting away from that heavy burden this time:

> This time Oscar didn't cry when they drove him back to the canefields. [. . .] The smell of the ripening cane was unforgettable, and there was a moon, a beautiful full moon, and Clives begged the men to spare Oscar, but they laughed. [. . .] They drove past a bus stop and for a second Oscar imagined he saw his whole family getting on a guagua,

even his poor dead abuelo and his poor dead abuela, and who was driving the bus but the Mongoose, and who is the cobrador but the Man
Without a Face, but it was nothing but a final fantasy, gone as soon as
he blinked, and when the car stopped, Oscar sent telepathic messages
to his mom (I love you, señora), to his tío (Quit, tío, and live), to Lola
(I'm sorry it happened; I will always love you), to all the women he had
ever loved—Olga, Maritza, Ana, Jenni, Nataly, and all the other ones
whose names he'd never known—and of course to Ybón. (320–21)

FINAL REMARKS

As in most trauma narratives, eventually fukú seems to have had the upper
hand in this story, too. Several reviewers have remarked that a *deus ex
machina* makes its appearance as the novel approaches its dispiriting end
(Owuor 2007; Scott 2007), which clearly connects Oscar's emotional
DNA with that of his victimised relatives and the harrowing history of the
country as a whole. Thompson and other social psychology theorists have
argued that, for the trauma process to get going, a connection needs to be
established between some fundamental injury inflicted on a community's
social norms and values and the victim's pain and mourning for the loss
(1998, 20–23). According to this analyst, it is only the "spiral of signification" that results from the "moral panic" of perceiving the direct line connecting past atrocities and present-day symptoms that can guarantee some
symbolic healing and reconstitution of the community: "Suffering has to
be witnessed to recognize its truth and injustice. The social recovery of
victims involves changing the threshold of moral vision both nationally and
internationally" (Humphrey 2002, 144). In this regard, the responsibility
that Junot Díaz has taken upon himself seems much larger than just trying to appease the ghosts that still haunt and weigh down the Dominican
diaspora, as he is also making readers in faraway nations around the globe
learn to empathise with and to transform their perception of their "traumatized condition" through art (see Bennett 2005, 6–11).

As the narrator of Díaz's novel explains, "anytime a fukú reared its
many heads there was only one way to prevent disaster from coiling around
you, only one surefire counterspell that would keep you and your family
safe" (6–7). In Yunior's opinion, this is just a simple word: "Zafa." But,
of course, in his case, we are aware that it has taken a little more than just
uttering that word and crossing his index fingers to make the curse and the
anguishing symptoms vanish. As Herman and other trauma scholars have
rightly observed, "the physioneurosis induced by terror can apparently be
reversed through words" (1992, 185). It is not surprising, then, that, when
he begins to put down Oscar's and his family's nightmarish history, Yunior
should be intrigued by the thought that "even now as I write these words I
wonder if this book ain't a zafa of sorts. My very own counterspell" (7).

NOTES

1. From *The Brief Wondrous Life of Oscar Wao by Junot Díaz*, copyright © 2007 by Junot Díaz. Used by permission of Riverhead Books, an imprint of Penguin Group (USA) LLC.
2. Caruth notes that, despite the unavailability of the traumatic event to consciousness, it intrudes repeatedly and belatedly into the lives of survivors and witnesses in the form of nightmares, flashbacks, hallucinations, ghostly presences, and similar manifestations (1996, 92).
3. For a fairly thorough discussion and assessment of these modes of representation, see LaCapra's essay "Writing History, Writing Trauma" in the volume included on the Works-Cited list.
4. Joseba Achotegui's work on the various types of grief and sorrows that invade migrants when they move to other countries may prove useful here. He coined the well-known term of "the Ulysses Syndrome" to describe some of these psycho-pathologies (2002).
5. Other Caribbean authors that could be included in this group of "hyphenated writers" are Cristina Garcia, Esmeralda Santiago, Piri Thomas, Edwidge Danticat, Oscar Hijuelos, and Julia Alvarez.
6. For an enlightening discussion of this "totalizing process" as it has happened in numerous post-colonial countries, see Chapters 3 and 4 of Homi Bhabha's *The Location of Culture*. Bhabha looks here into the dangerously ambivalent character of colonial discourse and the stereotypes it gave rise to.
7. Hayden White has discussed in several articles both the epistemological and ethical difficulties in carrying out this task (1999). In essays such as "The Problem of Truth in Historical Representation" and "The Modernist Event," he dwells upon some of the problems posed by a number of profoundly traumatic twentieth-century episodes to be represented in narrative.
8. To my knowledge, not much has been done so far in trying to figure out how gender issues may condition trauma processes. Díaz's novel would be an invaluable source of materials regarding this topic.

WORKS CITED

Achotegui, Joseba. 2002. *La depresión en los inmigrantes. Una perspectiva transcultural*. Barcelona: Editorial Mayo.

Adair-Hodges, Erin. 2008. "Ilk and Cookies: A Brief and Wondrous Interview with Junot Díaz." *Alibi Weekly* 17 (38), September 18–24. Accessed October 23 2008. http://alibi.com/index.php?story=24618&scn=art.

Alexander, Jeffrey C. 1987. "What is Theory?" In *Twenty Lectures: Sociological Theory since World War II*, 1–21. New York: Columbia University Press.

Alexander, Jeffrey C., Ron Eyerman, Bernhard Giesen, Neil J. Smelser, and Piotr Sztompka. 2004. *Cultural Trauma and Collective Identity*. Berkeley, L.A., and London: University of California Press.

Antze, Paul and Michael Lambek. 1996. *Tense Past: Cultural Essays in Trauma and Memory*. New York and London: Routledge.

Bennett, Jill. 2005. *Empathic Vision: Affect, Trauma, and Contemporary Art*. Stanford, California: Stanford University Press.

Bhabha, Homi K. 1994. *The Location of Culture*. London and New York: Routledge.

Caruth, Cathy. 1995. "Introduction." In *Trauma: Explorations in Memory*. Edited by Cathy Caruth, 3–12. Baltimore: The Johns Hopkins University Press.

————. 1996. *Unclaimed Experience: Trauma, Narrative, and History*. Baltimore and London: The Johns Hopkins University Press.

Díaz, Junot. (2007) 2008. *The Brief Wondrous Life of Oscar Wao*. New York: Riverhead Books.

Douglass, Ana, and Thomas A. Vogler, eds. 2003. "Introduction." In *Witness and Memory: The Discourse of Trauma*, 1–54. New York and London: Routledge.

Erikson, Kai. 1976. *Everything in its Path: Destruction of Buffalo Creek*. New York: Simon and Schuster.

Hanna, Monica. 2010. "'Reassembling the Fragments': Battling Historiographies, Caribbean Discourse, and Nerd Genres in Junot Díaz's *The Brief Wondrous Life of Oscar Wao*." *Callaloo* 33 (2): 498–520.

Herman, Judith L. 1992. *Trauma and Recovery: The Aftermath of Violence—from Domestic Abuse to Political Terror*. New York: Basic Books.

Humphrey, Michael. 2002. *The Politics of Atrocity and Reconciliation: From Terror to Trauma*. London and New York: Routledge.

Jelin, Elisabeth. 2003. *State Repression and the Labors of Memory*. Minneapolis: University of Minnesota Press.

Kakutani, Michiko. 2007. "Travails of an Outcast." Review of *The Brief Wondrous Life of Oscar Wao*, by Junot Díaz. *The New York Times*, September 4. Accessed October 28 2010. http://www.nytimes.com/2007/09/04/books/04diaz.html.

LaCapra, Dominick. 2001. *Writing History, Writing Trauma*. Baltimore and London: The Johns Hopkins University Press.

Lacan, Jacques. (1973) 1977. *The Four Fundamental Concepts of Psychoanalysis: The Seminar of Jacques Lacan, Book XI*. Edited by Jacques-Alain Miller and translated by A. Sheridan. New York: Norton.

Mansbach, Adam. 2007. "World of 'Wao' Contains Multitudes." Review of *The Brief Wondrous Life of Oscar Wao*, by Junot Díaz. *The Boston Globe*, September 16. Accessed October 28 2010. http://www.boston.com/ae/books/articles/2007/09/16/world_of_wao_contains_multitudes

Méndez, Danny. 2008. "In Zones of Contact (combat): Dominican Narratives of Migration and Displacements in the United States and Puerto Rico." PhD Diss., Austin: The University of Texas.

Owuor, Elisabeth. 2007. "A Dominican Teen Shadowboxes with his Past." Review of *The Brief Wondrous Life of Oscar Wao*, by Junot Díaz. *The Christian Science Monitor*, September 11. Accessed October 28 2010. http://www.csmonitor.com/2007/0911/p16s01-bogn.html.

Sacks, Samuel. 2008. "The Long Puzzling Absence of Junot Díaz." Review of *The Brief Wondrous Life of Oscar Wao*, by Junot Díaz. *Open Letters Monthly*, February 2008. Accessed October 28 2010. http://www.openlettersmonthly.com/junot-diaz.

Scott, A.O. 2007. "Dreaming in Spanglish." Review of *The Brief Wondrous Life of Oscar Wao*, by Junot Díaz. *The New York Times*, September 30. Accessed October 28 2010. http://www.nytimes.com/2007/09/30/books/review/Scott-t.html.

Sztompka, Piotr. 1993. *The Sociology of Social Change*. Oxford and Malden, MA: Blackwell.

Thompson, Kenneth. 1998. *Moral Panic*. London: Routledge.

Weich, David. 2007. "Junot Díaz out of the Silence." *Powell's*, August 17. Accessed October 23 2008. http://www.powells.com/authors/junotdiaz.html.

White, Hayden. 1999. *Figural Realism: Studies in the Mimesis Effect*. Baltimore and London: The Johns Hopkins University Press.

Part III
Trauma and the Problem of Representation

9 H. D.'s Twice-(Un)Told Tale

Marc Amfreville

As the oblique reference to Nathaniel Hawthorne invites us to remember, American texts were often published in one medium—journals, magazines— later to somewhat paradoxically start a new life in another form. Matters are somewhat more complicated as far as Hilda Doolittle's *Tribute to Freud* is concerned. The woman Imagist poet and novelist whose name Ezra Pound, once her fiancé, shortened to her initials, H. D., went to Vienna in 1933–34 to undergo psychoanalytical treatment with S. Freud. In London, in the autumn of 1944, she composed "Writing in the Wall," which was originally published in a magazine, *Life and Letters Today*, by instalments between 1945 and 1946. It constitutes the first part of the volume later published as *Tribute to Freud*. This text is explicitly presented by its author as a way for her to "remember," although there has been some misunderstanding as to *what* she wished to remember. To put it in a nutshell, although it appeared belatedly in *The International Journal of Psychoanalysis* in 1956, Ernest Jones' review set the tone and H. D.'s volume is chiefly considered as "the most enchanting ornament of all the Freudian biographical literature." Further likened to a "lovely flower [that] the crude pen of a scientist hesitates to profane" (H. D. 1984, vi), H. D.'s account thus appears as a refined, almost urbane, document on Freud himself, rather than the result of a creative process that powerfully transmuted an essentially *oral* experience into a *written* two-stage text that can in no way be only seen as a record. As Norman Holmes Pearson suggestively puts it in the preface: "remembering Freud was significant, for remembering him was remembering what she had remembered with him" (H. D. 1984, v). This chapter lays the emphasis on the uniquely intimate content of "Writing on the Wall," but should not lead us to overlook the fact that it is also a literary adventure, an attempt at writing what only lends itself with difficulty to transcription. After briefly recalling the circumstances of composition, thus trying to define the unique nature of this unclassifiable text, and endeavouring to demonstrate that it mimics the structure of trauma, we shall move on to give an elaborate example of the way it can be used, delving into the said and the unsaid, to inform us, through specific literary means, about the very nature of trauma.

* * *

What makes the literary endeavour clearly distinct from a mere record of psychoanalytical sessions is first and foremost the time that had elapsed since the cure—about ten years—and the juxtaposition of another text, the journal H. D had kept at the time of her sessions with Professor Freud. It is most relevant to underscore that the author herself thought it necessary to specify that "Writing on the Wall" was composed "in the autumn of 1944, with no reference to the Vienna notebooks of Spring 1933" (H. D. 1984, xiv). This title refers to the interplay of light and shadow that H. D contemplated at length from her bed on the Greek island of Corfu, a series of pictures that partook in the nature of dreams in a two-fold direction: the projected revelation of unconscious truths, but also the foreshadowing of future events, as perceived only by seers and poets. It is of course most interesting that the Imagist poet should think of these elusive pictures as "written," which goes to show that H. D. was already fully aware of the indelible nature of unconscious traces and of the intimate relation they bear to creation. In a letter to her lover Bryher, H. D., no doubt rightly—if somewhat irreverently, as often in her private correspondence about the "Master"—noted that Freud had been extremely interested in that double dimension: "Papa has nearly chewed off his right whisker with excitement over Corfu" (March 18 1933, quoted in Stanford Friedman 1990, 291).

The Vienna notes, we are told, had remained in Switzerland and she had no access to them until after the war, that is after having published "Writing on the Wall." With her typical sense of humour and sensitivity to language itself, H. D. goes on to present "Advent"—the title that she gave to her notes and that in itself summons up Christian associations to a "before actual birth" period, which a poet born in Bethlehem (Pennsylvania) to an Evangelical Protestant[1] family was bound to have in mind—as "the continuation of 'Writing on the Wall' *or its prelude* [. . .] taken direct from the old notebooks of 1933, though it was not assembled until December, 1948, Lausanne" (H. D. 1984, xiv).

There may be more to this hesitation between sequel and origin than meets the eye. At least it is the contention of this chapter that, with a degree of consciousness impossible to determine, H. D. thus chose an organisation for her work that closely mimics the structure of trauma. It is tempting to see the juxtaposition of the two terms, and the double emergence of salient elements of H. D.'s personal history as they reveal themselves in the successive accounts, as offering invaluable insight into the relationship between oral and written, far removed from a transcription. It is most significant in that respect that this double account thus powerfully blends two different moments and leads to the general collapse of the "narrating I" and "narrated I," or as psychoanalysis-inspired critique often has it, that of the "I now" with the "I then." As Susan Stanford Friedman convincingly summarises:

> H.D.'s texts about her analysis with Freud split the autobiographical
> subject, to construct an "I then" who was engaged in the talking cure

with Freud and an "I now" who repeats that initial experience as a writing cure. Each text doubles the analysis by recreating the primary scene of analysis in the past and then establishing a secondary scene of analysis constituted in and through the act of writing. (Stanford Friedman 1990, 293)

Everybody who has worked on trauma, and the inscription of trauma in a written form—fiction or non-fiction—readily remembers how Freud, in *Beyond the Pleasure Principle*, clarified his own views on the matter by resorting to a rapid summary of Tasso's *Jerusalem Delivered*. Let us then only recall briefly that Tancred unwittingly kills his fiancée disguised in the armour of an enemy knight and then kills her again by slashing a tree in which the young woman's soul has taken shelter. It may be worth remembering that Freud grounds the whole theory of what goes "beyond the pleasure principle" in this idea of a compulsive repetition, strikingly expressed in the improbable expression, "kills her again":

> In the light of such observations as these, drawn from the behaviour during transference and from the fate of human beings, we may venture to make the assumption that there really exists in psychic life a repetition-compulsion, which goes beyond the pleasure-principle. We shall now also feel disposed to relate to this compelling force the dreams of shock-patients and the play-impulse in children. (Freud, 1953–74, vol. 18, 16–17)

One is tempted to underscore the obvious since our critical ear is so accustomed to the title of Freud's essay that one tends to forget that what he really endeavours to describe and understand is the nature of an impulse that drives to commit, and to repeat, actions that our pleasure-seeking nature should lead us to steer clear of. Significantly enough, he thus resorts to the examples of the recurring nightmares of shock-patients—by which he means soldiers traumatised during World War I—and to the necessity for children to endlessly repeat the same games: two cases that, as we shall see, are closely related to Hilda Doolittle's attempt at mastering events or realities that could not be readily assimilated.

As suggested by Cathy Caruth, there is however something more specific to this story by Tasso and its use by Freud than the founding of trauma theory on the idea that trauma is always second, a point that Jean Cournut, a French psychoanalyst, had summarised in very convincing terms:

> A trauma always comes in the spurs of a preceding one: one can even wonder if the former may not have induced the following one, or even the chain repetition of the following ones. However, one calls "trauma" an intra-psychic perturbation that is not necessarily apparent and that often remains unconscious. The most violent event is not the most

> spectacular. Then, consequences are often just as misleading, covered
> up, buried, repressed, denied. Displacement, condensation, secondary
> elaboration, multiple screen plays just as in a dream: the acknowledged
> trauma is never the right one, a trauma always hides another one.
> (Cournut 1988, 18; my translation)

Cathy Caruth, particularly sensitive to the fact that Clorinda's voice is
heard bemoaning her second death at the hand of her beloved, helps us
take matters further:

> [trauma] is always the story of a wound that cries out, that addresses
> us in the attempt to tell us of a reality or truth that is not otherwise
> available. This truth in its delayed appearance and its belated address,
> cannot be linked only to what is known, but also to what remains
> unknown in our very actions and our language. (Caruth 1996, 4)

Informing these remarks is the whole subterranean theory of *Nachträglichkeit*, questionably translated as "deferred action," as the concept clearly
refers not to the action of Trauma Number One upon Trauma Number
Two, but the exact reverse, that is the way in which Trauma Number Two
sends the subject back to the scene of Trauma Number 1, in other words
provokes the reading of a past that precisely had been written but had
remained unread until Trauma Number 2. The fact that Tancred does not
hear the voice of Clorinda until the second wounding should thus not be
confused with a haunting (unless one is ready to consider the bidirectional
nature of haunting in literary terms), in a way that resembles Paul Auster's
definition of "rhyme" ("if these two events were to be considered separated
there would be little to say about either one of them. The rhyme they create
alters the reality of each" [Auster 1982, 161]).[2]

I would add that H. D.'s method in *Tribute* is, as always, palimpsestic.
(She did write a long text called *Palimpsest* after all.) Here the second text
("Writing on the Wall") is written over the first ("Advent"); the former
tends to pre-empt our interpretation of any material that may be found
in the latter. At the same time (so to speak), the central event of "Writing on the Wall" (the Corfu episode) pre-dates the meeting with Freud. In
other words, although "Advent" seems to be the matrix for "Writing on the
Wall", the reverse relationship is also present. The structure of the diptych
mimics the working of remembrance and oblivion, each cancelling out the
other: in this respect, it certainly offers a structural parallel with trauma.

This juxtaposition of sources—from psychoanalytical theory and the
realm of literature—helps us point to the essentially amphibian nature of
writings such as *Tribute to Freud*. To put it in rather dramatic terms: there
is no such thing as the transcription of a cure. Everybody familiar with the
psychoanalytical experience will readily acknowledge the utter impossibility of recording the volatile nature of the analysand's monologue. This is

not a poetic euphemism that would tend to point to the essentially mysterious nature of speech in these circumstances. Rather, it is a way to acknowledge the flimsy, uninteresting, matter-of-fact nature of the rubble in which gold nuggets are by definition quite rare. Even when H. D. is taking notes of her sessions, against Freud's repeated advice—it meddles with the psychoanalyst's injunction "not to prepare" her sessions, and will result in the interruption of the "transcription," which covers only a period from March to June 1933—, she obviously shows little interest in catching the oral quality of her enterprise. There are of course some examples of attempts at such a restitution. The very first sentence thus runs:

> I CRIED TOO HARD . . . went to the old wooden restaurant with the paintings, like the pictures that my mother did, Swiss scenes, mountains, chalet halfway up a hill, torrent under a bridge. (H. D. 1984, 115)

One notices the paratactic juxtaposition of the first words, all in block capitals, with the rest of the sentence which, in spite of the lack of a subject pronoun, runs almost like normal prose. The same is true of reported speech. While in many instances, H. D. allegedly reproduces the words of her analyst by merely quoting them, with due punctuation (e.g.: "He said, 'I doubt if even *you* could do that'" [H. D. 1984, 175, with italicised "you"]), she also frequently resorts to indirect speech, for more theoretical interventions: "The Professor said I had not made the conventional transference from mother to father, as is usual with a girl at adolescence" (H. D. 1984, 136). One may doubt the reality of such a comment that sounds almost like a passage taken from a development H. D. could have read in one of Freud's articles on female homosexuality. She describes herself as having free access to her analyst's personal library, and as being encouraged even to borrow books from him so as to spend her time usefully in Vienna between two sessions, one hour a day, five days a week.

It is unfortunately impossible to assess H. D.'s actual intention when she decides to include these odd seventy pages in her volume. We can however perceive them, like Clorinda's voice that had not been heard during the first stabbing, as *a crying out*. Even if H. D. insists that they were taken "direct" from her notebooks, we are bound to suspect that some manner of rewriting did take place. She even encourages our suspicion by specifying in the above-quoted note on the text that they were "assembled" in 1948—a word that clearly smacks of reviewing and editing. We can accept her word for it though, that *most of* the revisions in question were made at the moment of composition, that is presumably, on the very day, or at most one or two days at most, after each session (some undated entries do not allow greater precision on that point). What strikes the reader is not so much the actual oral quality of the document, as the way in which it apparently respects on many occasions the *logic of free association*. While the long passages to which Ernest Jones was particularly sensitive, that is the moments at which Freud's behaviour, interventions or surroundings are described, those that actually concern dreams or childhood memory bear an unmistakable stamp of truth. This does not mean that one cannot find interest in some of Freud's most personal and unexpected cues. For example: "'And I must tell you (you were frank with me and I will be frank with you) I do not like to be the mother in transference—it always surprises me and shocks me a little. I feel so very masculine'" (H. D. 1984, 146–47), a sentence so little in keeping with more recent analysts' entrenched silence, and (understandable) refusal to tell their personal secrets. But beyond that immediate almost frivolous voyeuristic seduction of the reader, the interest of H. D.'s enterprise lies elsewhere. Speaking of a tentative book of memories about Greece, she writes:

> I have tried to write of these experiences. In fact it is the fear of losing them, forgetting them, or just giving them up as neurotic fantasies, residue of the war, confinement and the epidemic, that drives me on to begin again and again a fresh outline of the novel. (H. D. 1984, 153)

The exact same thing could be said of the account of her sessions. The fear of losing not only the retrieved traumatic memories themselves, certainly wrenched from oblivion by the session itself, is also the very process that led to their discovery. It is perhaps the most striking aspect of this strange text that, precisely, nothing is brought to the surface that was not already known to the patient. Or rather, to be more specific, nothing is presented as having been a particularly shattering discovery in the way many cure accounts have accustomed us to read. Part of this is of course linked to the fact that what could have been considered as the disclosure of hitherto repressed material had already been revealed in part I. But this of course is a remark based on the reader's experience, as opposed to that of the writer—who should logically have consigned her discoveries. The only possible explanation is that H. D. unearthed relatively little buried traumatic material during those six months. It is then particularly challenging to have to fill the gap between what is still unknown in "Advent" and already integrated in "Writing on the Wall" to such a degree that it does not even require an explanation. Here again we are left with two conflicting hypotheses. H. D. may have wanted, when writing her *Tribute* in its final form, to leave out the most intimate aspects of her discoveries in order to privilege her presentation of the father of psychoanalysis. As Susan Stanford Friedman suggests, we can "interpret those traces and reconstruct what H.D. has resisted telling us. *Tribute to Freud* in particular represses the extent of her conflict with Freud, as well as the frank discussions they had about gender and sexuality. *Tribute to Freud*, in other words, replicated the 'refusal to speak' that Freud identified with the resistance" (Susan S. Friedman 1990, 298).[3] But then, the whole concept of resistance, in Freudian terms, cannot be seen as a mere conscious refusal to speak or write; it is linked with repression (in its strongest psychoanalytical sense) and the very secret of neurosis. H. D. then may, just as plausibly, not have made the findings that a reader, with access to both texts, is more likely to achieve. Such an interpretation receives the inkling of a confirmation when H. D retrospectively writes:

> The war closed on us before I had time to sort out, relive, and reassemble the singular series of events and dreams that belonged in historical time to 1914–19 period. I wanted to dig down and dig out, root out my personal weeds, strengthen my purpose, reaffirm my beliefs, canalize my energy, and I seized on the unexpected chance of working with Professor Freud himself. (H. D. 1984, 91)

The interest of the second possibility for our understanding of the text in particular and trauma studies in general is obvious. Following Freud's lead as playfully summarised by Jean Cournut, we would then be able to experience the fact that a trauma, like a train, can always hide another one.

* * *

The scope of this paper does not allow for complete development of this intuition, but we could perhaps take one telling example: that of H. D.'s brother, who died in France during Word War I. In that respect, it should be underscored that the interval between the two juxtaposed texts spans the period when Word War II was a constant fear in the author's mind—a fear that takes the form of extreme worries over the Professor being directly threatened by anti-Semitic Nazi sympathisers—to the moment when the war was almost over.[4] In that respect, it is not surprising that H. D. should have rapidly been led to speak of her brother, whose death constituted what she would have described as one of her haunting traumas. It should be recalled that the loss had been made doubly painful, as it had immediately brought about their father's heart attack and subsequent death. The crucial importance of that brother, while his death has not been evoked yet in "Advent," is brought to the reader's attention by Freud himself, who, commenting upon a dream, asks H. D. if she was Miriam (as opposed to being Moses himself in his basket, as a projected desire to be the founder of a new religion duplicated in the choice of "Advent" as a title . . .). We then read: "But the Professor insisted that I wanted to be Moses" (H. D. 1984, 120). No further comment is made. The reader begins to think that this Egyptian dream and Freud's intervention point to the fact that H. D. wanted to *be* the baby brother, universally recognised in her family as more important than she was. This easy diagnosis sees itself confirmed by H. D. describing "the dead baby" on Gustave Doré's illustration of "The Judgement of Solomon." Needless to say, the baby who appears on the picture, about to be cut into two, *is not dead* . . . A later dream of two boys in a cathedral (H. D had an elder brother and a younger one), in which the small one appears "de trop" (in excess) does not lead her to question her death-wish concerning her sibling either. A childhood memory staging H. D's baby brother crawling on the floor—provoking his sister's immense feeling of her superiority—brings about the mother's loving partiality for the little "dog" (that also accounts for H. D.'s desperately wanting to gain the love of Freud's chow rummaging about the couch while being secretly irritated by its presence). This might just be a banal enough case of elder sister's jealousy and would only with difficulty be related to trauma if H. D. did not gradually reveal her passionate attachment to her mother, a fixation that should have resolved itself, as mentioned in the text, but with apparent casualness, if the adolescent girl had turned towards her father later. Only now do we understand Freud's comment on being the Mother in transference, and only now do we begin to suspect that Hilda's brother's death was only the second trauma in a chain that started with the birth of their younger sibling. Right before interrupting her note taking at Freud's request, H. D. quickly reports a dream that confirms the centrality of the topic: "I had had a dream of the sea, fear . . . and this connected with my youngest brother, who had been the baby" (H. D. 1984, 185). We are then led chronologically to the beginning of "Writing on the Wall", and discover that H. D. is absolutely fascinated by the patient who immediately precedes her, and who, we learn from the start, soon dies in a plane crash in which he was the pilot. When he dies, Hilda rushes to Vienna to tell Freud how sorry she is, and the Professor's comment: "You have come to take his place" (H. D. 1984, 6) only becomes clear for the reader some 180 pages later, that is in the already-written and commented upon description of the Miriam/Moses' dream. The reader is left alone to understand what the precedence of the pilot in Freud's time table and history of his patients hides: although he could have unconsciously stood for the elder brother, H. D. started seeing him as her younger brother the day when he had asked her to swap hours with her and thus come to be with the Professor—with their mother in transference?—before he did. Is not H. D. telling us more than she herself realises when she forcefully states: "I do not want to become involved in the strictly historical sequence" (H. D. 1984, 14). Such a blurring of time—again, very typical of the structure of trauma and its oral or written accounts—is further confirmed by H. D.'s constant referring to "her" brother without specifying which of the two she means. Hence the crucial importance of a reconstituted childhood memory, most probably staging the elder one, which she concludes by saying: "she likes my brother better. If I stay with my brother, become part almost of my brother, perhaps I can get nearer to *her*" (H. D. 1984, 33).

However striking the chain of shocks later received, nowhere do we get the impression of a hidden trauma as much as in the following quotation:

> The news of the death of my father, following the death in action of my brother in France, came to me when I was alone outside London in the early spring of that bad influenza winter of 1919. I myself was

waiting for my second child—I had lost the first in 1915, from shock and repercussions of news broken to me in rather brutal fashion. (H. D. 1984, 40)[5]

This is later established by a whole development on the eponymous "pictures on the wall," interpreted by Freud as "a desire for union with my mother" (H. D. 1984, 44). At various moments in the rest of the account, we do feel that H. D is about to theorise matters, to interpret her fascination with a swordless Greek statue, to explain her harping on her brother in Cain-like terms ("Am I my brother's keeper?" [H. D. 1984, 101]), to relate the various theoretical insights she reported to her various passions for women, or finally to elucidate her relation to the Professor in the light of a repeated fear that he might "not take her"? She never does, and thus leaves her reader free to design his/her own way through her different accounts, and come to terms with her puzzling and stimulating creation.

* * *

Underlying this essay is the fairly recent direction taken by psychoanalytical reflections on trauma to better integrate Freud's concept of *Hilflosigkeit* (helplessness), and Ferenczi's specific contributions to clinical research on the subject. Thierry Bokanowski thus explains:

> And deprivations of love owing to the ignorance of a child's specific needs engender a 'psychic' sideration, due for the most part to despair. These effractions—symbolically equivalent to a psychic rape—rape of thoughts and affects by the disqualification of affect and the refusal on the part of the object (the mother or the environment) to acknowledge the child's feelings in a situation of distress—lead to a sideration of the self, asphyxia, and even the slow death of psychic life causing extremely painful states. (Bokanowski 2005, 31; my translation)

This insight—whose relevance is related to H. D.'s being very young when her brother was born (four years between the birth of the two brothers)—casts a decisive light on H. D.'s *Tribute* that *tells* us twice over the impossibility of *telling* what might be called the "Ur-trauma", perhaps precisely because the traces it has left belong to the realm of the infra-verbal.

One is then inclined to suggest that both oral and written attempts at going beyond trauma are subsumed in the very absence that signals its presence and inform its impracticable representation.

My reading of *Tribute to Freud*, clearly based on the individual realm of trauma, should not however lead to underestimating the importance of its historical and social dimension. While H. D.'s engagement with psychoanalysis is clearly rooted in personal ground and the desire to make sense of her most intimate experiences, she clearly remains conscious that part of her difficulties were shared by a whole generation of individuals who had undergone two world wars in such a short period of time. Furthermore, the constant anti-Semitic threat of Nazi terror that Freud and all Austrian Jews were living under at the time of her sessions—though the Professor remained oddly calm under the menace—must have played a decisive part in reactivating what H. D. herself called her "war phobia". H. D. repeatedly claims that her entire life and destiny were shaped by the two World Wars, and many of her books deal with the parallels she establishes between the two. Here is a telling extract from an autobiographical text entitled "H. D. by Delia Alton" (1949), in which she summarises and links together her various books before she gives a final statement as to the place of psychoanalysis in her personal evolution:

Certainly *The Gift*, the story of the Child, synthesises or harmonises with the Sigmund Freud notes. I assembled *The Gift* during the early war-years, but without the analysis and the illuminating doctrine or philosophy of Sigmund Freud, I would hardly have found the clue or the bridge between the child-life, the memories of the peaceful Bethlehem and the orgy of destruction, later to be witnessed and lived through, in London. That outer threat and constant reminder of death, drove me inward, almost forced me to compensate, by memories of another world, an actual world where there had been security and comfort. But this was no mechanical intellectual trick of mind or memory, the Child actually returns to that world, she lives actually in those reconstructed scenes, or she watches them like a moving-picture.[6]

In the process of doubling—Delia Alton is one of H. D.'s frequently used personae—H. D. objectifies, so to speak, the very course of poetic alienation that allowed her to move from the experience of psychoanalysis to the literary recreation of the memories of her therapeutic sessions and those of her childhood. Bearing in mind that Freud himself considered poets as his and our "masters,"[7] and that he repeatedly invited H. D. to consider herself special as such, we should perhaps retain the idea that, as psychoanalyst and literary scholar Liliane Abensour challengingly recalls in her "Trance and Poetic Transcription":

[I]n the analytical relation, if the analysand can, outside the timeless time of analysis see him/herself as the writer of a text that writes itself without ever writing itself, he/she can only be a poet outside the text of psychoanalysis. (1981, 206) [8]

NOTES

1. There is also a "Moravian" streak—both in the sense of the mystic Protestant brotherhood, and of the geographical place, that unites Hilda Doolittle to Freud, who was born in Moravia, which she dutifully noted. See H. D. 1984, 33.

2. For further analysis of the theoretical insights displayed by literature into the mechanisms of trauma and its relation to rhyme, see Marc Amfreville 2009).

3. In the same line of interpretation, as Professor Avril Horner kindly pointed out during the written exchanges that we had after a very stimulating question she asked at the oral presentation of my paper, various critics have underscored Hilda Doolittle's later critical engagement with psychoanalysis. See for example: Rachel Connor, *H. D. and the Image* (Manchester University Press, 2004): "Describing her relationship with Freud in *Tribute to Freud*, H. D. compares the investigation of her psychoanalytic sessions to the scientific observation of her male relatives (her father was a professor of astronomy and her maternal grandfather a microbotanist), feeling herself positioned between the 'double lenses' of her father's telescope and her grandfather's microscope. As an analysand of Freud, H. D. joins that group of women (Freud's patients, friends and peers) who were objects, not of the microscope or the camera, but of another kind of 'gaze', the scrutiny of scientific theory. Freud's methods often subjected women to an examination which monopolised their own theoretical reasoning, fixing them into a certain mould with his 'penetrative gaze and colonizing intelligence'; put simply, Freud's theory 'turned all women who deviated from the [heterosexual] model, or who self-assertively strove for equal rights, into suitable cases for treatment'" (10–11).

4. Let us recall that Freud died in London in 1939, having been rescued from occupied Austria by Marie Bonaparte, Princess of Greece, and a disciple of Freud herself, who, apart from that bail out, spent some of her immense fortune on the development of psychoanalysis in France.

5. One can only express surprise at the fact that no further comment is made throughout *Tribute to Freud* on the death of that first child, an essential link in the chain of traumas described or alluded to in the two sections of the book, which goes from childhood to World War II. Likewise, while her homosexual liaison with Bryher is made clear but not explicitly discussed, one is left to note that very little is said about the decay of her relation with and divorce from her husband, Aldington, nor about the fact that she almost died while giving birth to her daughter Perdita in 1919.

6. This autobiographical text is unpublished. I would like to give special thanks to my colleague and friend Antoine Cazé, professor of American Poetry, for generously sharing with me the notes he took on this material during his research trip to Yale in 2008 and for offering precious advice during the elaboration of this paper. (Beinecke Rare Books Library, Yale University, H. D. Archives, Box n° 44, folder 11.25).

7. "Story-tellers are valuable allies, and their testimony is to be rated high, for they usually know many things between heaven and earth that our academic wisdom does not even dream of. In psychic knowledge, indeed, they are far ahead of us, ordinary people, because they draw from sources that we have not yet made accessible for science." S. Freud 1953–74, vol. 9: 6).

8. This essay is dedicated to the memory of Liliane Abensour.

WORKS CITED

Abensour, Liliane. 1981. "Transe et transcription poétique." *Dire, Nouvelle Revue de Psychanalyse* XXIII : 11–17.

Amfreville, Marc. 2009. *Ecrits en souffrance*. Paris: Michel Houdiard Editeur.

Auster, Paul. 1982. *The Invention of Solitude*. New York: Penguin.

Bokanowsky, Thierry. 2005. "Le concept de trauma chez S. Ferenczi." In *Le Traumatisme Psychique*. Monographies de Psychanalyse. Edited by Michèle Emmanuelli Brette Françoise and Georges Pragier, 27–42. Paris: Presses Universitaries de France.

Caruth, Cathy. 1996. *Unclaimed Experience: Trauma, Narrative and History*. Baltimore: The Johns Hopkins University Press.

Connor, Rachel, *H.D. and the Image*. 2004. Manchester: Manchester University Press.

Cournut, Jean. 1988. "Du bon usage du trauma." In *Trauma réel, trauma psychique, Les Cahiers de l'Institut de Psycho-pathologie clinique*: 11–25 (my translation).

Doolittle, Hilda ("H. D"). 1984. *Tribute to Freud*. New York: New Directions.

Freud, Sigmund. *Beyond the Pleasure Principle*, vol. 18; *Delusion and Dreams in Jensen's Gradiva*, vol. 9. In *The Standard Edition of the Complete Psychological Works of Sigmund Freud*. 1953–74. London: Hogarth Press.

Stanford Friedman, Susan. *Penelope's Web*. 1990. New York: Cambridge University Press.

10 "Time to Write them Off"?

Impossible Voices and the Problem of Representing Trauma in *The Virgin Suicides*[*]

Bilyana Vanyova Kostova

Jeffrey Eugenides' straightforward choice to represent the multicultural U.S. society through a selection of "impossible voices" (Foer 2002, 78) has proved to be both an ambitious and successful enterprise. His two novels, *The Virgin Suicides* (1993) and *Middlesex* (2002), are shaped according to their author by Modernism (Eugenides 2004; Schiff 2006) but in fact they also draw on the multifaceted reality of Postmodernism, especially on issues concerning the problematising unity of meaning and integrity of the self, and most prominently of all, adolescent trauma. Ever since its revival in the late 1980s and early 1990s, Trauma Studies has been approached from the point of view of the modern era and with the notable concern for individual and collective psychic health in times of wars, economic depressions, technological takeover, and other disasters. In the search for a remedy for this "wound of the mind," as Freud defined it in 1920 using the Greek conception of the term, story-telling has been pointed out by trauma theorists as a therapeutic device because of its ability to transmit, in Dominic LaCapra's words, "a plausible 'feel' for experience" (2001, 13). Not only is narrative meant to arouse empathy in the readers but it also serves as a means to work through traumatic experiences and their memories, wrapping them in meaning, and helping victims of trauma to cope with the overwhelming events from their past and their haunting consequences. Eugenides' debut book, *The Virgin Suicides*, quickly became a cult novel due to the juxtaposition of a rare first person plural narrator with an exaggeratedly marked need to review the past—which plausibly conveys the abovementioned "feel" for traumatic experience—and an uncanny topic which lurks behind the discursive presence of five suicidal sisters, the central and obsessive focus of the collective narrator. Although later in his career the author became more and more hostile towards first person narrators, at the time he tried to "make it new," in his own words, by exploiting the collective voice of such first person narrator (Eugenides 2004).

The story of the novel is set in a dreamy, magical realist Detroit suburb in the 1970s, characterised throughout the book by Motown music, the decline of the auto industry and the ecological deterioration of the neighbourhood. There, among the many foreign surnames of its residents, the would-be collective narrator lives: an indistinct group of boys who years later, as grown-up men, undertake a quest to unravel the truth surrounding the enigmatic suicides of their neighbours—the five Lisbon girls. This narrator strives with a need for accuracy to report on both the lives of the neighbour sisters and the disintegrating post-war suburbia life they once enjoyed. In this attempt, they create a collective memory out of a number of flagrant contradictions and fragments. Furthermore, another locus of interest is the melancholic tone of the grown-up narrator. Interspersed with moments of grim comic relief and teenage misunderstandings, the tone invests the fragmented narrative with a magical atmosphere that takes on now a gothic, now a detective, or now a romantic turn. From an early stage, this collective voice was compared to Faulkner's "A Rose for Emily"—a tale included in a collection of love stories edited by Eugenides—by calling it "a direct homage to Faulkner's use of the first person plural" (Shostak 2009, 810). The author has dissented from this opinion, pointing to another source of literary heritage in the person of the Nobel-Prize winner Gabriel García Márquez and specifically the we-narrator in his renowned novel *The Autumn of the Patriarch* (1975) (Schiff 2006, 105–106).

Not only has García Márquez inspired Eugenides in terms of narrative voice and themes but, as we shall see, he has also left a magic realist imprint on his first and charismatic novel. Its narrative continuously folds back on itself, gradually disclosing the fatalistic thirteen months of the Lisbon girls' tragedy while at the same time it compiles reminiscences of tragic loss, adolescent desire and vivid traumatic events that are textually recaptured in the form of interviews, stolen objects and photographs. Depicting and interpreting such events requires a specific mode of narration, one which can both "transform the mundane into the extraordinary," as a *New York Times* reviewer claimed (Berne 1993), and also confer realistic tinges on these extraordinary stories. For this reason, Eugenides chooses to rely on a hybrid mode made famous by García Márquez, magical realism, a mode that characteristically goes beyond our common notions of the fantastic and the factual in a matter-of-fact tone, i.e. without causing "a perceptible break in the narrator's or characters' consciousness" (Benito, Manzanas and Simal 2009, 43).

The present chapter explores the way the collective narrator perceives and endeavours to recount the "unrepresentable" from their adult perspective as former witnesses of the suicidal events that happened when they were still teenagers. Joseph Flanagan accurately notices that "traumatic memory already takes the aesthetic modality of postmodernism" for two reasons: that it "is more faithful than normal memory because it does not distort that experience through representation" and that "a postmodern aesthetic is a more 'faithful' rendition of history" (2002, 390). After twenty years of bearing the unrepresentable testimony and acting it out through repetitive group gatherings, the boys, now turned into ageing men, decide to "write it off" or in other words, to find a language for trauma, which in Cathy Caruth's words is the "paradoxical obligation to speak without burying the silence at the heart of the story" (2006, 2). The language that results from the narrator's post-traumatic condition shuts out itself in the above-mentioned magic realist mode, and fluctuates between assuaging the real experience by situating the story in a safe and remote past in order to come to terms with it, and denying it an objective reconstruction of the events. The final target of the therapeutic inscription is just as paradoxical as the effects of the narrator's mental condition: to heal themselves from the trauma experienced as "empathic unsettlement" that this shocking experience has unleashed in their adolescence and, simultaneously, to maintain a structural and thematic "fidelity to trauma" (LaCapra 2001, 22).

As some critics and the writer himself have pointed out, the very title and opening paragraph of *The Virgin Suicides* seem to spoil every expectation of suspense by announcing what Caruth denominates "the enigmatic core" of the narrative plot (1995, 5). The five chapters in the book are moulded around the inexplicable first attempt and later successful suicide of the psychically disturbed youngest sister Cecilia Lisbon, two events that agitate the whole suburban area and set the boys' voyeuristic eyes permanently and obsessively on the Lisbon residence, and more specifically on the five daughters Cecilia, Lux, Bonnie, Mary and Therese, as one after another they end up also committing suicide. The riddle that these now middle-aged men are trying to solve twenty years after the "domestic tragedies" occurred, lies not so much in the chronological succession of events but in the mysterious cause of the unexpected suicides, which arouses the collective narrator's own uneasiness and sense of guilt deriving from these events. Therefore, it is not at all surprising that the starting point of their investigation of the girls' motives should be the last suicide, committed by Mary, without offering any clues to explain the earlier ones. The strategy of announcing a demise early in the plot has been expertly exploited by the Nobel-prize winner Gabriel García Márquez in his short fiction ("Big Mama's Funeral," "Death Constant beyond Love") and in his acclaimed novel *Chronicle of a Death Foretold* (1981). Eugenides himself acknowledged having read the Colombian writer's most famous novel, *One Hundred Years of Solitude* (Foer 2002; Schiff 2006), but, as Francisco Collado argues (2005), it is from *Chronicle of a Death Foretold* that he borrows both the overall structure based on interviews with witnesses and also themes such as memory, the circularity of time, uncertainty and the magical representation of reality so characteristic of García Márquez's fiction. The very beginning of the "long longing," as the author calls *The Virgin Suicides* (Weich 2002), tackles all these topics through the contaminated focalisation of the narrator's collective figure, which adopts the imaginative interpretations of their boyhood wherever their objective research fails.

The first traumatic event related in the initial paragraph of the story matches the one from the closing chapter, that is to say, it initiates the line of events which leads the collective voice to narrate the successive suicides—starting with Cecilia's slitting of her wrists—but it is also the event with which the Lisbon tragedy ends, marking the encompassing or circular strategy the telling of the story possesses.[1] Additionally, the first sentence in the book already serves as an introduction to an everyday reality rendered magical, a mode that has been suggested even earlier, in the title of the novel, by combining the innocent, even religious term "virgin" with its semantically antagonistic "suicides," suggestive of violence and sinful death:

> On the morning the last Lisbon daughter took her turn at suicide—it was Mary this time, and sleeping pills, like Therese—the two paramedics arrived at the house knowing exactly where the knife drawer was, and the oven, and the beam in the basement from which it was possible to tie a rope [. . .] Cecilia, the youngest, only thirteen, had gone first . . . (1993, 3)

Considered separately, neither the suicidal pact of the sisters nor the household atmosphere of a seemingly peaceful suburban area near Detroit would mean an essential departure from a possible historical representation of the 1970s, since it was an era of suburban movement, characterised by the decay of the downtown areas and the consequent massive flight to the suburbs, a shift already under way following the end of the Second World War. But it was also a time socially marked by a rise in youth suicide and by the collective suicides promoted by doomsday cults such as the Peoples' Temple.

Already in the 1950s and 1960s, the notion of living in the middle-class suburbs was attacked by some critics for failing to live out the American dream of individualism, now demolished by "the tyranny of the happy work team" (Whyte 2002, 401) or for serving "as an asylum for the preservation of illusion" (Mumford 1961, 494). Eugenides' view combines these classical critiques with William Mann Dobriner's concept of the heterogeneous nature of the suburbs, where the escape from class, race and religion is only apparent (1963). Despite being systematically avoided, African-American residents are depicted throughout the novel as the marginalised class of suburbia, whose improved living conditions are seen as a threat: "nothing shocked us more than the sight of a black person shopping on Kercheval" (99). Moreover, the novelist underlines in his story the common concurrence reported by critics on the issue of the contemporary return in the suburban areas to church and religion on a massive scale. But whereas this alleged moral upswing was understood to be more a secular than a sacred one (Berger 1968, 41), the title of the novel does actually hint at the threat which religion may pose, especially when taken to its extremes as is the case of the strictly Catholic matriarch Mrs Lisbon who, in trying fanatically to instil and reinforce a religious lifestyle in her daughters, ends up by oppressing and suffocating them. The result of such brainwashing is exemplified in the image of "the laminated picture of the Virgin Mary [Cecilia] held against her budding chest" (4) while the young girl is killing herself in the bathtub. Despite this apparent attack on religious dogma, the collective narrator uses precisely religious terms to picture the girls, especially sexually deprived Lux, who is depicted towards the end of the novel as a kind of statue of Mary Magdalene, and her suicidal sister Cecilia, whose death is later on seen as the result of unquestionable fate:

> Under the molting trees and above the blazing, overexposed grass those four figures paused in a tableau: the two slaves offering the victim to the altar (lifting the stretcher into the truck), the priestess brandishing a torch (waving the flannel nightgown), and the drugged *virgin rising up* on her elbows, with an *otherworldly smile* on her pale lips. (1993, 6; emphasis added)

Lux's love encounters are rendered in a very similar light: she is described as a "carnal angel" with "two great beating wings" whose "eyes shone, burned, intent on her mission as only a creature with no doubts as to either Creation's glory or its meaninglessness could be" (148). Additionally, the unrecognizable boys she sleeps with become "footholds in Lux's ascent," perplexed at her "measureless charity," a description which resembles the position of the would-be narrator—the voyeuristic boys—as pawns in the girls' suicidal plans (212). Such terms, however, do not reinforce a religious discourse but instead sound like a parodic spell that softens harsh witnessing experiences, such as Cecilia's attempt to die by her hand or Lux's sex marathon on the roof of her own house. In close proximity to such descriptions, there are humorous passages that emotionally detach the readers from the religious and attach them to the worldly, such as the grossness of the paramedics or the boys' imaginative inventions of love techniques; they act as criticisms of religious repression. Ironically, the two most atrocious Christian sins—lust and suicide—coexist with the very symbols of Christianity: the laminated pictures of the Virgin Mary, harbingers of suicidal messages, and the crucifix "draped with a brassiere" (9), described with a language of traumatic paradox: a hesitant quivering between lyrical magic realistic scenes interrupted by comic intermissions. The engulfing language of magic realism, on the one hand, shuts out the disturbing settings like a protective screen while the comic sections reveal the gross materiality of love and death by drawing the reader's attention to the irony they often entail. For instance, on the night of the girls' mass suicide, sultry Lux is almost canonised as a saint: "For a second it [Lux's face] didn't seem alive: it was too white, the cheeks too perfectly carved, the arched eyebrows painted on, the full lips made of wax" (210), converting her transiently into a Jungian anima. The next moment she is unzipping Chase Buell's trousers claiming "you guys have got me all worked up" (211). These subtle ironic comments and incongruities bring together the inability of religion to ease the private tragedy of the Lisbon family with the larger scale suburban problems of Detroit in the late 1950s, and the decline of their auto industry, and the appearance of immigrant delinquency. To these growing social problems could be added the ever incessant preoccupation with environmental problems from the early 1960s on, which led to the publication of books such as Rachel Carson's *Silent Spring* (1962)—a volume that raised consciousness on the detrimental influence of man on nature—or John Fuller's *We Almost Lost Detroit* (1975)—a book that deals with the nuclear meltdown in Detroit's breeder reactor in 1966.

It is when the setting and related situations have been established as plausible, that the unpredictable combination can take place, or to use H. H. Arnason's words (1968, 363), "the fantastic juxtaposition of elements or events that do not normally belong together," which creates the magic aura from an early stage of the narrative. The novelist juxtaposes the idea of five much loved sisters living in their "comfortable suburban home" (1993, 5) with their inexplicable self-inflicted violence. The progressive attempts to solve this riddle come in the literary combination of obsessive romance, eerie imaginary and impossible reality offered by magical realism. The fantastic component never really takes full shape—Cecilia's ghost turns out to be her living sister, the contagion of suicide described in García Márquez's style appears more parodic than credible ("Household objects lost meaning," 158), the creepy smell that engulfs the Lisbon household originates either from the untidy house or from the contaminated lake. However, the mode transmits the sense of a distant, uncertain, barely possible reality. Additionally, the girls' beauty is caricatured as grotesque and marvellous: they have two extra canine teeth (63), Bonnie seems "a foot taller than any of her sisters, mostly because of the length of her neck," Therese possesses "the cheeks and eyes of a cow," Mary has a chipped tooth and a moustache (42), Lux laughs gruffly, Cecilia stares uncannily with her yellow sunflower eyes. Magical realism also springs from the fairy-tale descriptions of suburban lush settings ("ivy-covered house on the lake"), immaculate lawns, shrubbery, "heaps of trees throwing themselves into the air" (34) or of the doomed Lisbons' house with a cloud always hovering over the roof (141).

The use of magic realistic motifs in the novel is twofold: they provide a way to deal with trauma and represent the "unrepresentable," and they can be envisioned as an articulation of a socio-political critique towards the U.S. suburban ethos that disguises the haunting presence of black population, unhealed post-war traumas, immigrant issues and economic depressions. They lend themselves perfectly to the mythic atmosphere of the setting and false sense of community that living in suburbia entails. The narrator's effort to denounce the normalisation of the five deaths by the community is triggered by the "empathic unsettlement," or the urge to "write them off," to bury the girls figuratively by producing in LaCapra's words, a record of "socially engaged memory work" based on the combination of imaginative memory and realistic fantasy (2001, 66). There is certainly something to be denounced about the conflicting interaction between the supernatural quality of the events and the plausibly depicted hypocritical society at Grosse Point: the suburban fallacy, the American *unmelting* pot, and the environmental destruction, and all this is achieved without detriment to the lyrical quality of the story because, as Theo D'Haen argues, magical realism is also "a means for writers coming from privileged centre of literature to dissociate themselves from their own discourses of power, and to speak on behalf of the ex-centric and un-privileged" (1995, 195)—in our case, the young girls.

Going back to the beginning of the novel, a mass suicide against a domestic background on the very first page of the book might anticipate the reader's lack of interest in a story whose end has already been disclosed, but the shock of bereavement it causes to the collective narrator keeps the reader's interest throughout the scattered pieces of dubious information. From the perspective provided by Trauma Studies, it becomes clear that magical-realist and experimental stylistic effects can be associated with LaCapra's concept of "empathic unsettlement" (2001, 41). The narrator's deep personal concern becomes an important element that is reflected in the style in which the bare facts about the five premature deaths are presented. Out of the few certain facts they have at their disposal, the collective voice tries to create "a story [they] could live with" (241), a narrative construct told from the distant temporal position of the grown-ups but which establishes a bond with the dead girls. In the narrative of the events, the boys' early focalisation of the suicides, years later expressed as magical realism, appears to be the result of an empathic response when the suicides took place, an empathy that eventually demands a coherent explanation for their traumatic loss.

However, this does not mean that the story's report unfolds following a fluent narrative path: in addition to setting up the beginning *in ultimas res* and suddenly jumping to Cecilia's failed first attempt to commit suicide, the narrative continues to go back and forth in time, pausing every now and then for the examination of a photograph or another random token that belonged to the girls, without eschewing the ebullient figure of speech, typical of trauma: the hyperbole. To these experimental strategies, the author adds the fragmentary condition of the layout within chapters, with blank spaces here and there, a device that indicates the traumatised narrator's fragmentary memories of the events and also the conscious but difficult pull to work through trauma by offering a coherent representation of the past. Towards the end of the first such fragment in Chapter 1, the collective narrator makes explicit their role in the course of reporting evidence gathered from the neighbours and while pondering over a photograph of the house:

> We've tried to arrange the photographs chronologically, though the passage of so many years has made it difficult. A few are fuzzy but

revealing nonetheless. Exhibit #1 shows the Lisbon house shortly before Cecilia's suicide attempt [. . .]. As the snapshot shows, the slate roof had not yet begun to shed its shingles, the porch was still visible above the bushes, and the windows were not yet held together with strips of masking tape. (1993, 4–5)

As was the case in the previous paragraph, in which an elderly neighbour is interviewed about the day before Cecilia's fatal decision, the men that make up the collective narrator would have the reader believe that their investigation is truthful and reliable in its mimicking the processes of memory and forgetfulness which take place with the passing of time. Therefore, it seems inevitable that they should supply the missing pieces with a patchwork of different discourses: the witnesses' confusing depositions, the narrator's own dreamlike interpretations of pictures, objects and songs obtained from the girls, and finally the girls' own silent cryptic messages. But the effect achieved is the result not only of the jumbling of so many diverse piece, but also of the narrator's inclination to distort the traditional process of narrativisation by introducing a sense of anachronicity by means of prolepses and analepses, for instance, that of the collapsing house in the quotation above; these abound in the novel and, according to Gérard Genette, are typical of homodiegetic narratives due to their retrospective character (1980, 63). This sense of anachrony, combined with the hyperbolic lyrical descriptions, insist on the delay and incompletion of knowledge, and originate in the narrator's twenty-year-long acting out. This repetitive acting out is manifested in the pointless gatherings of the boys at which they repeatedly went through their collection of fading trifles, enigmatic reminiscences of the girls: "rather than consign the girls to oblivion, we gathered their possessions once more, everything we'd gotten hold of during our strange curatorship" (186).

Additionally, the collective narrator's traumatised condition after their direct witnessing of the events is explained with reference to some noticeable intertextual features of the book. Nabokov's *Lolita* (1955), identified early on as one of Eugenides' literary influences in *The Virgin Suicides* (Griffith 1994), also features going through pictures—"I am going to pass around in a minute some lovely, glossy-blue picture-postcards" (1980, 9)—as well as exhibits—"Exhibit Number Two is a pocket diary bound in black imitation leather . . ." (40)—and random objects belonging to the young girl—"I cherished and adored, and stained with my kisses and merman tears, a pair of old sneakers, a boy's shirt she had worn, some ancient blue jeans . . ." (253). Although these intermissions do not break the narrative flow as excessively as happens in *The Virgin Suicides*, Nabokov's autodiegetic narrators like Humbert Humbert (*Lolita*) or Victor ("Spring in Fialta") certainly share with the collective voice in Eugenides' novel the capacity to remember their obsessive desire for the adolescent girls, and formulate a "reenactment of the literary process by which we fall victim to, and memorialize, our loves," as Eugenides has commented (Wright 2008).

Gradually readers become trapped in a setting where a number of long-term witnesses, in love with the girls, become the traumatised makers of an uncertain narrative, trapped in the prison of their fragmented and incomplete memories where magical realism and journalistic factuality clash and fuse. Even photograph descriptions, which are meant to bring about a sense of representational veracity, fail in their purpose. They play the role of palpable proofs for the narratee, a figure which is addressed explicitly more than once: "Please don't touch. We are going to put the picture back to its envelope" (1993, 119). The pictures, however, are not textually present in the book; they are only described and interpreted by the narrator, therefore failing to fulfil their mission of transparency. All interviews and reports, on the other hand, apparently help to mend the holes in the narrator's story left by the wrong encoding of the dramatic events: "you may read it for yourself if you like; we've included it as Exhibit #9" (95). Be it photographs or other types of evidence, these "objective" items of research and restoration of the past acquire as much importance as the narrator's subjective affect towards their dead teenage loves, resulting in a continuous epistemic clash.

Despite their contradictory nature, we might think that the narrator's reports also represent in the novel the dominant male Western empiricist impulse to rationalise and examine scientifically the evidence so as to extrapolate valid results from it. However, while the structure mirrors the rational-scientific worldview, the story itself goes in another direction: that of the preindustrial mythical realm represented by the suburb, the new refuge away from the "impoverished city" with its factories and slums (34). The chaotic use of time is more apparent than real. Everything in myth narratives is organised so that it moves towards entelechy or perfection: deep beneath the superficial chronological confusion there emerges a circular pattern and symmetry in the plot: the narrative is brought to a close not twelve, as might be expected, but thirteen months after the youngest sister's unsuccessful attempt, because of Mary's failure to put an end to her life. Thus, the failed attempts and suicides are balanced towards the beginning and towards the end of the narrative. Even though the men as collective narrator remember living in a "timeless murk" during the year of the suicides (141), where time is marked by the seasonal fish fly plagues or leaf-raking ventures, "years later" they are already able to distinguish between their childhood myth-making mentality and the analytically gathered information provided by external witnesses. For example, the second chapter retells Cecilia's suicide apparently as it has been registered by witnesses around the neighbourhood, in this way adding new details and simultaneously confronting them with the information provided by the collective narrator earlier in the first chapter. According to Debra Shostak, this "dual temporality" underlines the boys' effort to present themselves as "objective gatherers of information" (2009, 816). However, besides these two tenses, there is one more timeline that sneaks into the emotional narrative: the present tense that is used every now and then to reveal the timelessness of the sisters' presence in the adult narrator's everyday life. Either on examining Cecilia's diary (she "hadn't thought about any of us. Nor did she think about herself. The diary *is* an unusual document . . ." 42; emphasis added) or by considering their own bodies as substitutes for the girls' ("even though that winter was certainly not a happy one, little more *can* be averred. Trying to locate the girls' exact pain *is* like the self-examination doctors *urge* us to make" 170; emphasis added), the narrator slips into this almost ungraspable shift in tenses—from past to present—which delineates the persistent presence of the absence that is so frequently dealt with in trauma narratives. Despite the years past, the incomplete memories still haunt the survivors, a condition that becomes represented by the strategic devices of blurring the temporal boundaries and making leaps back and forth in time, producing the impression of what Genette terms anachrony. LaCapra suggests that "in post-traumatic situations in which one relives (or acts out) the past, distinctions tend to collapse, including the crucial distinction between then and now" (2001, 46). In this way, the collective voice becomes stuck in the mythic "timeless murk" of the boys' adolescence and their search for an exit out of it in their attempt to make things clear by means of producing a narrative. However, the chronological puzzlement between past and present in the examples above is indicative of the processes of acting out trauma where the boys confuse the girls' identities, writing and bodies with their own. For instance, on reading Cecilia's diary the teenage boys experience vicariously the state of being a girl: "after one of us had read a long portion of the diary out loud [. . .] [w]e felt the imprisonment of being a girl" (43).

As mentioned earlier, the temporal dislocation apart, the text has been given an enthusiastic reception by some critics because of its curious, experimental narrator, which has been defined in this chapter up to now as a "collective" entity, characterised by being a first-person-plural voice in the manner of a "Greek chorus." Eugenides affirms that he devised this strategy simply from the idea of "a group of boys" of undefined number, who observe, testify and fabulate the multiplicity of subjective truths into a striking testimony of the events that took place twenty years earlier (Miller 2002). Although the names of Chase Buell, Kevin Head, Joe Hill Conley, Paul Baldino or Tom Faheem are reiterated in the narrative, the members of this group are never fully portrayed, either psychologically or physically. A deeper insight into their individual essence is in great part prevented by the choice of the collective narrator and its *testimonial* functions. Significantly, whenever any small detail revealing their personality turns up, it unfailingly contributes to the characterisation of suburban life or of their particular generation:

> There had never been a funeral in our town before, at least not during
> our lifetimes. The majority of dying had happened during the Second

World War when we didn't exist and our fathers were impossibly skinny young men in black-and-white photographs [. . .] Their own parents, who spoke foreign languages and lived in converted attics like buzzards, had the finest medical care available and were threatening to live on until the next century. Nobody's grandfather had died, nobody's grandmother, nobody's parents, only a few dogs . . . (35)

In this paragraph, as in García Márquez's novel *Chronicle of a Death Foretold*, the boys are separate from but are also part of the whole communal involvement: the collective voice reinforces the impression of an unhealed wound deeply implanted in the whole neighbourhood. This contrasts with the individualistic middle class values in Western countries and consequently reinforces the community-oriented worldview frequently associated with magical realism and "primitive" cultures. However, irony springs from the fact that this modern community seems to be a nostalgic utopian recreation of an uncorrupted preindustrial milieu. Until the outbreak of the suicides, the boys participate in the general illusory scheme, ignorant and innocent of death, violence, and traumas because the society they belong to enshrouds problematic affairs like racism, post-war traumas, and environmental pollution, which nevertheless start to come to the surface when the boys grow up and gradually wise up in their collective role as narrator of the Lisbon sisters' tragedy. In this sense, then, the girls as direct victims and the boys as problematised witnesses epitomise the troubled condition of their whole community. The latter are regarded here as "problematised" because of their double status of direct witnesses of and later commentators on traumatic events, which explains the confusion of timelines and the presence of symptoms of acting out (the teenage boys' identification with and haunting by the girls) and of working through (their narrative will to understand and commemorate the events many years later).

The collective homodiegetic narrator acts for the most part as a bystander or narrator-as-witness, but occasionally swerves into a more overt autodiegetic direction, especially when any of the boys come into close contact with the girls. This narrative strategy unavoidably brings to mind Charles Dickens' Pip, whose adult self relates the story of himself as a little child, lending realism to every childish fancy. Here, too, the mature narrator, positioned outside the diegesis, focalises through the eyes of its dazzled adolescent personas and marks the moral, intellectual, and temporal distance with comic and grotesque comments. Paul Baldino, for instance, gains the other boys' respect by supplying the following beguiling information: "'I saw the movie,' he said. 'I know what it's about. Listen to this. When girls get to be about twelve or so'–he leaned toward us–'their tits bleed'" (11). Whereas the focalisers' function is producing a realism-enhancing effect by limiting their knowledge to what can be learned by observation and deduction from their childish perception of the events, the adult narrator makes the readers detect this unreliability and provides them with other realism-enhancing features like the distortion of memory as a consequence of the passing of time but also of the impact of trauma. However, the narrator is also quite self-conscious: the men that it is comprised of are aware of themselves as fiction makers, and allow for some metafictional and therefore realism-undermining effects to clash with their role as reporters who are searching for the truth.

According to Linda Hutcheon, postmodern ideology is premised on such paradoxical tension: it is simultaneously different from but connected with the past (1988, 125). In this sense, *The Virgin Suicides* longs to return to a traditionally told story where the final truth may be found—that is, where narrative may give coherence to the collective narrator's traumatised memories—but it also accentuates the fact that this return to a naïve realist type of representation is problematic if not impossible, and hence the experimental character of the whole narrative—which also indicates that the trauma is still being worked through. This tension between realistic reliability and improbable fabulation materialises in the figure of the collective narrator as impersonation of both testimonial credibility springing from the direct witnessing of suicide, and phantasmal instability springing from the magic realist mode of narrating. It could also be likened to what LaCapra calls "middle voice," that is, "the 'in-between' voice of undecidability and unavailability or radical ambivalence of clear cut positions" (2001, 19). Since the narrator in Eugenides' novel is both active—the ageing men perform as objective historians and researchers of the girls' lives—and passive—they are also traumatised victims of unprecedented circumstances—such a definition only stresses the postmodern preoccupation with the unavailability of the past, or to be more precise, the troubled relation between past and present in terms of memory, trauma, and representation. It is important to bear in mind that the collective narrator is influenced partially by a "discursive variant of the middle voice [which] may at times be linked to an ethos of uncertainty, risk, more or less indiscriminate generosity, and openness to the radically other, who is utterly unknown" (197). The unknown other might not only be the dead girls, but also the traumatic response that these girls have awakened in the narrator, a response which has given place to an uncertain and hesitant discursive remembrance. However, speaking in strict terms, it seems impossible to apply the Greek grammatical form of the middle voice, even if the author himself has Greek origins. An Anglo-Saxon language is not entirely equipped with the necessary linguistic structures that would allow for "avoiding the unwanted intrusion of the subject [the narrator] into the action it wants to represent" (Kellner 1994, 133). This centredness on action can also be regarded as a reflection of the unreliability, and undecidability proper to trauma.

Chase Buell, a member of the narrator, advises at one point: "They're [the girls] just memories now [. . .] Time to write them off" (186). This writing off is carried out once the narrator initiates an empathetic storytelling which not only helps in verbalising and (partially) giving sense to trauma but also in transmitting it to somebody else. An imaginative storytelling admits the double telling of which Cathy Caruth speaks: "the oscillation between a crisis of death and the correlative crisis of life: between the story of the unbearable nature of an event and the story of the unbearable nature of its survival" (1996, 7). Instead of putting the blame for the girls' suicides on the repressive power of their mother or on the traumatised narrator's own shattered encoding of the events, or on the doctors' tentative interpretations and the neighbours' conjectures, the narrative focuses on the very subjects who demand active remembering, emotional engagement, and "socially engaged memory work"—as I mentioned earlier. The narrator chooses to give voice to the five disempowered teenagers not only to produce a sustained and convincing critique of the North American suburb as a simulacrum of illusive happiness and suffocating banality: the novel is an attempt to dialogue with the "dead intimates," to use LaCapra's words, as a way of understanding them, of refusing to bury their memory, of leaving oneself open to being affected by it. It also expresses a need to close a long-lasting chapter of melancholic regression to the year when their childhood was interrupted. Not surprisingly, the lingering nostalgia in the last pages of the book stands for "fidelity to trauma" and paradoxically for the need to be free of compulsive acting out.

The confrontation with reality—the coming out of childhood by learning about erotic desire and death at the same time—exacerbates both the shock and distress in the young boys as internal focalisers when the events happened, and magnifies the belated response in their role as adult narrator. Even though twenty years have elapsed, the narrative reconstruction of the girls' and the boys' past is still mediated by the shared suffering of the collective narrator, explicitly manifested in the first person plural pronoun "we" and the magic realistic, almost sacralised, aura of the suicides. The narrator's exuberant language, far from playing down the importance of the tragic demises, which on a larger scale are supplements of the whole suburban demise, creates what LaCapra calls "a 'feel' for experience and emotion which may be difficult to arrive at through restricted documental methods" or even objective realistic descriptions (2001, 13). The collective narrator's is a quest for overcoming their own traumatised condition, for humane understanding and love, so scarce in the postmodern era marked by divisions, boundaries and fragmentation.

*The research carried out for the writing of this essay is part of a project financed by the Spanish Ministry of Economy and Competitiveness (MINECO) (code FFI2012–32719). The author is also grateful for the support of the Government of Aragón and the European Social Fund (ESF) (code H05).

NOTES

1. A recurrent device in García Márquez's fiction, especially in *One Hundred Years of Solitude*.

WORKS CITED

Arnason, H. Harvard. 1968. *History of Modern Art*. Ann Arbor: University of Michigan Press.

Benito, Jesús, Ana Mª Manzanas, and Begoña Simal. 2009. *Uncertain Mirrors: Magical Realisms in US Ethnic Literatures*. Amsterdam and New York: Rodopi.

Berger, Bennet M. 1968. *Working-class Suburb: A Study of Auto Workers in Suburbia*. Berkeley and Los Angeles: University of California Press.

Berne, Suzanne. 1993. "Taking Turns at Death." The New York Times. Accessed June 10, 2010. http://www.nytimes.com/1993/04/25/books/taking-turns-at-death.html.

Caruth, Cathy. 1995. *Trauma: Explorations in Memory*. Baltimore and London: The Johns Hopkins University Press.

———. 1996. *Unclaimed Experience*. The Johns Hopkins University Press.

———. 2006. "An Introduction to 'Trauma Memory and Testimony'." *ReadinOn* 1 (1): 1–3. Accessed June 10 2010. http://readingon.library.emory.edu/issue1/iss1toc.htm.

Collado Rodríguez, Francisco. 2005. "Back to Myth and Ethical Compromise: García Márquez's Traces on Jeffrey Eugenides's *The Virgin Suicides*." *Atlantis* 27 (2): 27–40.

D'Haen, Theo L. 1995. "Magical Realism and Postmodernism: Decentering Privileged Centers." In *Magic Realism: Theory, History, Community*. Edited by Lois Perkinson Zamora and Wendy B. Faris, 191–208. Durham: Duke University Press.

Dobriner, William Man. 1963. *Class in Suburbia*. New Jersey: Prentice Hall.

Eugenides, Jeffrey. (1993) 2002. *The Virgin Suicides*. London: Bloomsbury.

———. 2004. "Two Novelists on the Legacy of Joyce." *Slate*. Accessed February 5 2011. http://www.slate.com/id/2102446/entry/2102452.

Flanagan, Joseph. 2002. "The Seduction of History: Trauma, Re-Memory, and the Ethics of the Real." *CLIO* 31 (4): 387–402.

Foer, Jonathan Safran. 2002. "Eugenides: An Interview." *BOMB* (Fall): 75–80.

Genette, Gerard. (1972) 1980. *Narrative Discourse*. Translated by Jane E. Lewin. New York: Cornell University Press.

Griffith, Michael. 1994 "The Virgin Suicides." *The Southern Review* 30.2: 379–393.

Kellner, Hans. 1994. "'Never Again' is Now." *History and Theory* 33 (2): 127–44.

LaCapra, Dominick. 2001. *Writing History, Writing Trauma*. Baltimore: The Johns Hopkins University Press.

Miller, Laura. 2002. "Sex, Fate, and Zeus and Hera's Kinkiest Argument." *Salon*. Accessed July 10 2010. http://www.salon.com/2002/10/08/eugenides_3/

Mumford, Lewis. 1961. *The City in History: Its Origins, Its Transformations, and Its Prospects*. New York: Harcourt Brace.

Nabokov, Vladimir. (1955) 1997. *Lolita*. London: Penguin.

Schiff, James A. 2006. "A Conversation with Jeffrey Eugenides." *The Missouri Review* 29 (3): 100–19.

Shostak, Debra. 2009. "'A Story We Could Live With': Narrative Voice, the Reader, and Jeffrey Eugenides's *The Virgin Suicides.*" *Modern Fiction Studies* 55 (4): 808–32.

Weich, Dave. 2002. "Jeffrey Eugenides Has It Both Ways." *Powells.com.* Accessed August 12 2010. http://www.powells.com/blog/interviews/jeffrey-eugenides-has-it-both-ways-by-dave/..

Wright, Annabel. 2008. "Q and A with Jeffrey Eugenides." *5th Estate.* Accessed August 12 2010. http://www. fifthestate.co.uk/2008/01/q-and-a-with-jeffrey-eugenides/.

Whyte, William Hollingsworth. 2002. *The Organization Man.* Philadelphia: University of Pennsylvania Press.

11 Fugal Repetition and the Re-enactments of Trauma

Holocaust Representation in Paul Celan's "Deathfugue" and Cynthia Ozick's *The Shawl*

María Jesús Martínez-Alfaro

INTRODUCTION

Paul Celan—an anagram on Ancel, the Romanian spelling of his family name—was born in 1920 in Czernowitz, a German-speaking, formerly Austrian city which was annexed by Romania but whose Jewish population maintained their allegiance to German culture. In 1941 Romanian and Nazi troops plundered, murdered and deported thousands of Jews. In June 1942 the ghetto was searched and Celan's parents were caught. None of them survived the war. Celan remained in Romania, working in several labour camps until the Russian liberation of 1944. He eventually settled in Paris, married a Frenchwoman, and became a naturalised French citizen. He attained fame as a poet, but he had to battle severe depression. In 1970 he committed suicide, drowning himself in the Seine.

Celan's "Todesfuge," "Deathfugue" in English, is regarded by many as *the* canonical poem about the Holocaust in any language. "Deathfugue" describes a death camp experience, although the setting is unspecified, and the same goes for the first person plural which includes the poetic speaker in the poem. The opening lines (1–3) are repeated almost verbatim at the beginning of each of the following sections of the poem. Other lines are also repeated with slight variations, and these cyclical repetitions provide the poem with its encantatory as well as claustrophobic rhythm, the claustrophobia of being caught on the threshold of what is an almost certain death for the "we" of the poem. The initial metaphor reveals a world upside down, as milk, which is nourishing and life-giving, has turned black, fusing with the darkness of the experience that the poem tries to capture. The first person singular, the reader soon finds out, is that of the victims, camp inmates trapped in the oppressive life-in-death circles of everyday existence in the camp. There, a blue-eyed German commandant orders the Jews to dig their own graves while others have to play musical instruments, thus setting the scene for murder carried out to the accompaniment of music. He shoots the camp inmates dead and tells those still alive that they will become smoke rising up to the sky (l. 25) and that they will find their grave in the air (l. 4, l. 15, l. 33), thus referring to their corpses being burnt in the camp crematorium.[1] Death seems to be the only possible escape, "escape" being one of the meanings of the "fugue" in the title: "Todesfuge" is the "Fugue of Death." But, until death comes, if it does, the camp inmates are caught in the grip of endless suffering and agonising repetition.

The poem was originally entitled "Todestango." As John Felstiner points out, to call the poem "Todestango" was "to cancel out everything in life that would be svelte, graceful, nonchalant" (1986, 252). This reference to the tango, or to music, in the title also forged a macabre connection between the poem and true fact. Scholars and survivors have more than once commented on the presence of music in some camps. Thus, Michael Levine recounts how, in November 1943, over 18,000 Jewish men, women and children were shot by the S.S. in ditches near the crematorium in Maidanek camp as part of what the Nazis called "Operation Harvest Festival": an S.S. operation to exterminate the Jews in the Lublin district of Poland. As the killings were taking place, camp loudspeakers played the popular German tune "Rosamunde, Give Me a Kiss" (Levine 2006, 139). Felstiner refers to the same camp as the focus of a 1944 pamphlet, whose author—Konstantin Simonov—describes a mass death march towards the camp crematorium while loudspeakers emitted "the deafening strains of the foxtrot and the tango. And they blared all the morning, all the evening, and all night" (quoted in Felstiner 1986, 252). The pamphlet's title was "The Lublin Extermination Camp," that is, Maidanek camp, established on the Southern outskirts of Lublin, in Eastern Poland. It may well be that Celan even read this pamphlet before writing his poem.

Significantly, Lublin is also the second name of the main character in Cynthia Ozick's *The Shawl*, her first name being not Rosamunde, as in the popular German tune played at the Lublin camp, but, close enough, Rosa, Rosa Lublin. Just as in the Lublin camp music muffled the sounds of pain, Ozick's *The Shawl* is also about the silencing effects of and the traumas brought about by unbearable suffering springing from Holocaust experience.

The Shawl is made up of two stories which were published separately in the *New Yorker*: "The Shawl" (1980) and "Rosa" (1983). They were brought together in 1989 and published in book form with an epigraph taken from Celan's "Deathfugue." "The Shawl" elliptically portrays the extreme circumstances of life in the camps by focusing on a triangle of characters: Rosa Lublin, her fourteen-year-old niece Stella, and Rosa's baby child Magda. The story reaches its nadir as Rosa witnesses in silence how little Magda is killed by a camp officer. The second story, "Rosa," takes place almost forty years later and is an exploration of the main character's life as shattered by the traumas of the past. The events in the first story make up the second story, as these events, or rather the protagonist's searing memories of them, obsessively return in "Rosa."

In what follows, I will take the connection established by Ozick's epi-graph to *The Shawl* as a point of departure to delve, firstly, into the com-plexities inherent in the representation of the Holocaust and Holocaust trauma as illustrated by Celan's poem and Ozick's short stories. Then, I will analyse the way in which the re-enactments of traumatic experience are similarly given textual form in these works: in order to convey the effects of trauma and the protagonist's feeling of entrapment in/by the past, Ozick's short stories are marked by a dynamics of repetitions punc-tuated by slight variations, which can be likened, I contend, to the (fugue) dynamics at the core of "Deathfugue." Celan's poem is written in a con-tinuous present tense that somehow evokes the indelibility of trauma, its endless return and atemporality. *The Shawl*, on the other hand, definitely shows how trauma haunts the traumatised person's existence, making the past present, disrupting chronology and mental balance. Both Celan and Ozick express in this way the agony brought about by the unwanted repetition of deadly experience. Finally, I will focus on the possibility of working through, or at least lessening the effects of trauma, by address-ing the question with which these works confront the reader: the question of whether there is a way out, an opening, or whether, in Primo Levi's words, "the injury cannot be healed" (1989, 24). To answer this ques-tion, I will consider the role of the other, more specifically, the import that an empathic listener might have for the traumatised person whose only escape from the entrapment of trauma may be the existence of an "addressable you" that listens to his/her anguish.

THE LIMITS OF HOLOCAUST REPRESENTATION

Much has been written about the Holocaust and the limits of representa-tion, and, if one had to situate Ozick somewhere in this debate, it would be on the side of those who question the conjunction of fictional literature and the Holocaust. As Susanne Klingestein explains, the publication of *The Shawl* was

> a quiet literary sensation, not because a Shoah story was considered one of the best thirteen books of 1989 by the *New York Times Book Review*, but because the narrative seemed to violate a boundary that Cynthia Ozick's earlier misgivings about the nature of fiction had first helped to construct. She had alerted us to the danger inherent in fictional renditions of the Shoah, to an indecency which is partly caused by the absorption of the real into the imaginary and partly by the necessarily mythic quality of fictional renditions, since fiction thrives on narrative order and the creation of meaning. Precisely these two elements, however, were absent from the reality of the Shoah. (1992, 173)

Many of Ozick's works deal with the consequences of the Holocaust, but only in "The Shawl" has she focused on life in a concentration camp. Although both "The Shawl" and "Rosa" were written in 1977, it took several years for Ozick to make up her mind to publish them. Her extreme sensitiveness as to the moral quandaries of Holocaust literature is accountable for this long delay and hesitation. Holocaust representation is and will always remain problematic.

Paul Celan also immersed himself in the rough sea of Holocaust representation, feeling, after the publication of "Deathfugue," the dangers inherent in the poetic rendering of unspeakable horrors. This poem first appeared in a 1947 anthology, translated into Romanian. In Vienna 1948, he published his first book—*Der Sand aus den Urnen* [*The Sand from the Urns*]—including "Deathfugue" as the concluding section, and then, in 1952, the poem was published in Germany as part of the collection *Mohn und Gedächtnis* [*Poppy and Memory*]. "Deathfugue" became a national obsession in Germany, it was included in anthologies and textbooks, and hundreds of essays were written about it. However, what was studied and emphasised was the poem's aesthetics, its artfulness, its prosody and structure, its cadence and use of metaphor. To Celan's horror, the poem's real theme—the Nazi death camps and genocide—was, more often than not, ignored (Felstiner 1986, 253).

Something that particularly hurt Celan in the 1950s and 1960s, and surely influenced his evolution as a poet, was the fact that Theodor Adorno's famous dictum about the barbarism of writing poetry after Auschwitz was thought to be a veiled reference to "Deathfugue." There were even German critics who accused Celan of eliciting aesthetic pleasure from the Holocaust, presumably feeling their interpretation backed by the authority of Adorno's views. In the end, Celan turned against his poem, which he prevented from being reprinted in more anthologies. He even changed his writing style into a less explicit, less melodious, more fragmented and elliptical verse. As Felstiner explains, though, the truth is that Adorno had not heard of Celan when he wrote his essay "Kulturkritik und Gesellschaft."[2] Moreover, Adorno's editor Rolf Tiedemann told Felstiner that "it was Celan's poetry itself which led Adorno in 1966 to recant specifically his famous dictum" (Felstiner 1986, 255). Adorno returned to his statement about poetry and Auschwitz in a later essay and redefined its emphasis: while acknowledging that he had no wish to soften his verdict that to write poetry after Auschwitz was barbaric, he also admitted that literature had to resist this verdict, in other words, be such that its mere existence after Auschwitz proved that it had not surrendered to cynicism (Adorno 1982, 312).

In "Deathfugue," Celan was resisting Adorno's verdict even before its formulation and he went on doing so in his later poetry. Similarly, in *The Shawl*, Ozick also confronted the notion that it is barbaric to write poetry after Auschwitz not only by writing about the Holocaust, but also by writing lyrically, poetically, about it. To Ozick, writing about the Holocaust is always the result of a long internal debate, which forces her to face not only certain critical stances about the limits of Holocaust representation but also her own views, as a Jew and a Zionist, that certain forms of representation are antithetical to Judaism. In essays and interviews, she has repeatedly argued that invented fictional worlds are a breach of the Second Commandment, a form of idolatry, a re-enactment of paganism. Louis Harap highlights this as she explains that Ozick "interprets the Mosaic commandment against idolatry to mean not only to reject worship of material objects or images, but also not to pursue anything for its own sake apart from moral or religious status. Thus literature enjoyed for its own sake as an aesthetic object is 'idolatry'" (1987, 167). Ozick herself voiced her permanent inner struggle at a roundtable discussion during a conference held in 1987:

> Finally, about writing fiction. In theory, I'm with Theodor Adorno's famous dictum: after Auschwitz no more poetry. And yet, my writing has touched on the Holocaust again and again. I cannot *not* write about it. It rises up and claims my furies. All the same, I believe that the duty of our generation, so close to the events themselves, is to absorb the data, to learn what happened. I am not in favour of making fiction of the data, or of mythologizing or poeticizing it. [. . .] I constantly violate this tenet; my brother's blood cries out from the ground, and I am drawn and driven. (quoted in Horváth 1996, 257)

Literature—be it poetry or fictional narrative—can take us where history cannot. Paul Celan knew, and the same goes for Cynthia Ozick. The pull to write literature was, for both of them, as for many others, stronger than all dilemmas and misgivings. Making literature out of catastrophes of such magnitude as the Holocaust will time and again force authors and readers alike to confront the limits of representation. But, as Berel Lang suggests, such a confrontation will at least imply that we have rebelled against the most absolute of limits, the limit against which all representations must in the end be measured. And this limit, and thus the alternative, is silence (Lang 2000, 71).

REPETITION AND VARIATION: FUGUE DESIGN AND THE AFTER-EFFECTS OF TRAUMA

The title of Celan's poem points to the perspective from which to interpret the dynamics of the text, based, like the fugue, on repetition and dissonant counterpoint. A typically baroque form, the fugue consists essentially of a musical idea—a main subject melody called "theme"—that opens the piece and is then repeated several times in recognisable but slightly altered fashions. One or sometimes two themes are introduced at the beginning of the fugue, which are replayed and revisited cyclically in different voices and in various keys and rhythmic and tonal variations.

This dynamics of the fugue as explored and exploited by Celan in his poem can also be said to fit the workings of trauma and, more specifically, the way in which Ozick gives them narrative form in *The Shawl*. Initially, it is the epigraph that connects Ozick's work with Celan's poem. Yet the link goes beyond the Holocaust theme. The contrapuntal design of Celan's composition and its recurrent repetitions—intended to render the anguish of existence in extreme circumstances—are also at the core of *The Shawl*, which resorts to a similar conjunction of repetition with variation in order to reflect the way in which the subject who lives caught in the grip of trauma "unwittingly undergoes its ceaseless repetitions and reenactments" (Laub 1992, 69). Traumas re-emerge in different guises as uncanny repetitions of a past that remains present. In *The Shawl*, the traumatic experiences undergone by protagonist Rosa Lublin feed obsessions and fixations that trap her in the same oppressive circles of repetition that conform Celan's poem.

At its simplest a fugue begins with an exposition of the subject (theme) in one of the voices alone and in the tonic key. Then a second voice enters, repeating the subject in non-strict imitation. This so-called "answer" mimics the subject at a different pitch, often transposing it to another key, too. While the answer is being stated, the voice in which the subject was previously heard continues with new material, called "countersubject," which is also repeated by subsequent voices. The fugue's exposition concludes when all voices have stated or answered the subject. Fugues rarely stop after their initial exposition, which is often the first of three parts: exposition, development and recapitulation (New World Encyclopedia 2008). However, it is the dynamics of the fugue's exposition that most clearly accounts for the structure of Celan's poem.

"Deathfugue" begins with a statement of the theme—to use fugue terminology—built on the oxymoronic metaphor of the "black milk" and the image of obsessive drinking which has been explained as evoking the drunkenness of torture (Felman 1992, 30). The theme is stated in lines 1–3. Then this theme is repeated in non-strict imitation in lines 10–12, lines 19–21, and lines 27–29. These four sets of three lines constitute the beginning of the four main sections which make up the poem. Thus, the second, third and fourth parts start, like each of the voices in the fugue, with a restatement of the main subject in answer to its initial statement in the opening lines of the poem (ll. 1–3). After these opening lines, and as happens in the fugue's countersubject, more ideas are introduced (ll. 4–9). The following sections repeat them after the first three lines, repetition counterpointed by slight variation here as well (ll. 13–18, ll. 22–26, ll. 30–34). In this way, the universe of the poem emerges in front of the reader's eyes: the black milk drunk and drunk by the camp inmates; the blue-eyed German commandant; the inmates who are ordered to play music while others dig their graves; graves in the ground and graves in the air; the commandant's shouts, the lead bullets with which he shoots Jews dead, his hounds, his writing and daydreaming; Death as a master from Deutschland; darkness falling; Margareta and Shulamith.[3] The poem's universe thus expressed makes up a rich polyphonic composition which captivates the reader with its disquieting mixture of aesthetic beauty and stark horror, inviting him/ her to get lost in the maze of its elliptical lines, its repetitions, its marked rhythm, and its continuous present tense unhindered by punctuation.

***

As is the case with "Deathfugue," indeterminacy marks the beginning of Ozick's "The Shawl," the first of the two stories in *The Shawl*:

> Stella, cold, cold, the coldness of hell. How they walked on the roads together, Rosa with Magda curled up between sore breasts, Magda wound up in the shawl. Sometimes Stella carried Magda. But she was jealous of Magda [. . .]. Magda took Rosa's nipple, and Rosa never stopped walking, a walking cradle. There was not enough milk; sometimes Magda sucked air; then she screamed. Stella was ravenous. Her knees were tumours on sticks, her elbows chicken bones. (Ozick 2007, 3)

There is no reference whatsoever to the setting or to the characters' destination, though the reader soon discovers they are Jews marching towards a concentration camp. Permanently afraid of being discovered, Rosa hides her baby in a shawl. There is a focus on milk, as in Celan's poem. If milk has turned black there, here it is present through its lack. Again, something associated with life and nourishment turns into its opposite as Rosa's breasts soon run dry and produce no milk at all. If the child cries out of hunger, she will be discovered, and killed. Yet the baby relinquishes her mother's breasts quietly, takes the corner of the shawl, and milks it instead. Rosa thinks of it as a magic shawl, since it seems to keep Magda quiet and to nourish her with its "milk of linen" (7). She "sucked and sucked" (4), almost non-stop, like the inmates drinking their black milk in Celan's poem. The oxymoron in Celan's "black milk" is at the core of the shawl, whose white is stained by the ashes from the crematoria chimneys. Moreover, the shawl is connected with life and death: it keeps Magda alive but also brings about the child's murder.

One day Stella takes away the shawl because she was cold, and Magda walks into the square outside the barracks where she had remained hidden so far. Rosa sees her, howling "Maaaa . . . aaa!" (Ozick 2007, 8), her first and last word, "mama" in the child's tongue. The mother fetches the shawl, but it is too late: one of the guards has already caught Magda and is about to throw her onto the electric fence. This is how the moment is described, and this is also the most problematic passage of the story because of its poetic quality, which some have seen as a rather questionable aesthetisation of Holocaust horrors represented here by the cruel murder of a child:

> She [Rosa] stood for an instant at the margin of the arena. Sometimes, the electricity inside the fence would seem to hum; even Stella said it was only an imagining, but Rosa heard real sounds in the wire: grainy sad voices. [. . .] The voices told her to hold up the shawl, high; the voices told her to shake it, to whip with it, to unfurl it like a flag. Rosa lifted, shook, whipped, unfurled. Far off, very far, Magda leaned across her air-fed belly, reaching out with the rods of her arms. She was high up, elevated, riding someone's shoulder. [. . .] She looked like a butterfly touching a silver vine. And the moment Magda's feathered round head and her pencil legs and balloonish belly and zig zag arms splashed against the fence, the steel voices went mad in their growling, urging Rosa to run and run to the spot where Magda had fallen from her flight against the electrified fence; but of course Rosa did not obey them. (9–10)

One could resort to Celan's words in "Deathfugue" to say that Magda truly finds "a grave in the air," while time stops for the mother who witnesses in silence. What comes after this is a moment of paralysis that may be seen as affecting Rosa's life from this point on. She does not run towards her child because she knows that she will be shot if she does. Even if she lets her pain out and screams or cries, she will be shot. Thus, she stands still, watching, and stuffs the shawl inside her mouth: the shawl that kept Magda silent, and therefore alive, now ensures her mother's silence, and therefore, her own life, too.

The passage above illustrates the failure of human comprehension when the limits of what is bearable are brutally transgressed. The horror of what is taking place in front of Rosa's eyes is expressed in an indirect way, as if she needed a screen, as if her mind was unable to register things as they are really happening. Ozick manages to convey the brutality of the event by conveying the state of mind of the mother. She involves the reader in the account, demanding that s/he should decode the metaphors and imagine what is not said, thus showing, simultaneously, the necessity and the impossibility of portraying Holocaust horrors.

The lyrical description of Magda's death is a risky gesture and the result is as effective as it is disturbing. Yet there is also something disturbing about Magda's birth, or Magda's conception, to be more specific. Not much critical attention has been given to this point, which is more often than not mentioned in passing. From the very beginning Magda is described as an angelic creature, connected with air and beauty. Her complexion, the narrator explains, was not dark like Rosa's. Magda's eyes were so blue and her hair so blond that "you could think she was one of *their* babies" (Ozick 2007, 4). Stella looks at her during the march and exclaims "Aryan" (5), referring to the baby. Nothing more is said in the first story, but the seeds of suspicion are planted in the reader's mind.

"Rosa," the second story, gives the reader information about the time before and, above all, the time after the events recounted in the first story. We learn that Rosa's childhood was spent in the refined and genteel world of assimilated Polish Jewry. Her comfortable life came abruptly to an end as her family was forced to move to the Warsaw ghetto. Then she was taken to an unnamed concentration camp. After the war, she settled in New York and so did Stella, although the latter's strategy for survival has been to forget and this is one more reason accounting for the gulf that separates Rosa and her niece. Shortly before the narrative in "Rosa" begins, the protagonist, now nearing sixty, smashes her antiques shop in Brooklyn and moves to Miami, as is not unusual among many retired Jews. Stella sends Rosa money and she lives among the elderly in a shabby "hotel" (Ozick 2007, 13). Rosa's isolation is complete except for her recent habit of writing letters: to Stella, in crude immigrant English, and to Magda, in an elegant, literary Polish. She asks Stella to send her the shawl, which has the power to conjure Magda up and which stands for a traumatic past that Rosa obsessively acts out. It is a metonymy for the memory of Magda, which haunts Rosa as Holocaust memories haunt survivors.[4] It is in one of these letters, to Magda, that Rosa tackles the subject of her paternity:

> His name was Andrzej. Our families had status. [. . .] We were engaged to be married. We would have married. Stella's accusations are all Stella's own excretion. Your father was not a German. I was forced by a German, it's true, and more than once, but I was too sick to conceive. Stella [. . .] can't resist dreaming up a dirty sire for you, an S.S. man! Stella was with me the whole time, she knows just what I know. (42)

It would not be farfetched to think that a mentally unbalanced woman, who has created an alternative reality in which her daughter is not dead, may also be rejecting the truth about that daughter's conception to the point of believing that Magda's father was her boyfriend before the camps and not the S.S. man that raped her more than once. It is precisely because Stella was with Rosa all the time that she knows the unspeakable truth of Magda's paternity and blurts it out when she calls the baby "Aryan" in "The Shawl."

In this light, it could be argued that the (*ur-*)trauma affecting Rosa has to do with her rape, pregnancy and delivery in a concentration camp. Magda's murder is a second blow. The abstract and aestheticised language used to describe Magda's death conveys, above all, Rosa's mental flight from the scene. And just as the reader has to read through this screen, s/he also has to read through the screen memories that hide the truth about Magda's conception, a trauma which Rosa has repressed. As Dori Laub puts it, trauma may exert its power as a structuring and shaping force even in those cases in which it is not truly known to survivors, or even when silenced memories are not recognised by them as representing memories of trauma. It does nonetheless find "its way into their lives, unwittingly, through an uncanny repetition of events that duplicate [. . .] the traumatic past" (Laub 1992, 65). This is exactly what happens in *The Shawl*.

One of the most complex and interesting aspects of Ozick's work has to do with the linking of the traumas of birth and death in the two stories. Magda seems to come alive after the fact, precisely when she leaves the barracks and utters her first word, thus emerging from her long-standing muteness. The metaphor used to describe the event—"Magda's mouth was spilling a long viscous rope of clamor" (Ozick 2007, 8)—invites the reader to see that first word as an umbilical cord linking the child to the mother at the moment of birth, when the baby's cries break out and put an end to fearful silence. Yet this belated birth becomes one with the child's death, as she is discovered and killed. Such a confluence of birth and death is repeated at the moment of conception, a rape which is like death for the mother but that marks the beginning of life for the child. Rosa's physical immobility and mental paralysis as her child is being killed can be seen as repeating the paralysis that overtook her while she was being raped.[5] As Michael Levine insightfully points out:

> The moments of life and death are so densely intertwined here [in *The Shawl*] in large part because the child's life begins, in effect, with the unacknowledged death of Rosa, who was left pregnant and mortified by rape—pregnant, that is, not merely with a new life but with a virulent new strain of life-in-death. [. . .] The violent circumstances surrounding Magda's conception are never explicitly spelled out in the text [. . .]. Yet it is impossible to read *The Shawl* without taking this state of affairs into account. Not only do fragments of this traumatic scene continue to wash up on the shores of Miami and the edges of Rosa's consciousness forty years after the fact, but the rape is silently and unwittingly acted out in a number of guises and on a variety of stages throughout the text. (2006, 128–29)

Almost four decades later, in the second story, Rosa is still trapped between the trauma of her daughter's murder and that of her own rape. As has been pointed out, Rosa's physical immobility, mental paralysis and silenced screams are part of the account of Magda's death, but they powerfully contribute to making the scene appear as a *belated repetition* of the first trauma, that of the rape. Moreover, these two related traumas of the first story re-emerge again in the second story, and although they reappear several times and in different ways, I will concentrate here on a concrete episode that is particularly illuminating when it comes to illustrating the pattern of repetition with variation that figures among the after-effects of trauma, and that links its representation in *The Shawl* with the dynamics of the fugue in Celan's poem.

The episode begins in the laundromat near Rosa's hotel.[6] There, she "sat on a cracked wooden bench and watched the round porthole of the washing machine. Inside, the surf of detergent bubbles frothed and slapped her underwear against the pane" (Ozick 2007, 16–17). The setting and the situation are rather banal. And yet, there is a hint of violence connected with Rosa's underwear repeatedly beaten against the glass, which is what she focuses on, what she sees through the pane. Rosa watching as if mesmerised, something soft hitting something dark, the liquid element, all this is not far from some of the metaphors used to described Magda's death, when the child is seen "swimming through the air" before being "splashed against the fence" (9–10). This moment—Rosa watching the washing machine—somehow takes us back to Magda's murder, then, but it also sets in motion a thread in the story that will lead to a hallucinatory resurgence of the rape, that is, a repetition in a different guise of the sexual abuse inflicted on Rosa in the past.

While she is waiting for the washing machine to finish, an elderly man called Simon Persky tries to engage her in conversation. He is also a Jew from Warsaw, although he migrated to the States before the war. Rosa accepts his invitation to have a tea in a local cafeteria, but she soon returns to her hotel with her laundry. When she realises that she has lost a piece of underwear, her first thought is that she dropped it in the cafeteria and that "Persky had her underpants in her pockets. Oh, degrading. The shame. Pain in the loins. Burning. Bending in the cafeteria to pick up her pants" (Ozick 2007, 34). The middle section of this quotation seems to take the reader to another event, the buried trauma of her rape which erupts through the crevices of the present moment.

When the sun starts going down, Rosa sets out in search of her underpants. Her search leads her to the private beach of a hotel. The description of the setting, seen through Rosa's eyes, immediately brings the concentration camp onto the scene: the fence and its locked gate, the barbed-wire surrounding the beach, the towered hotel roofs in the distance compared to merciless teeth. . . . On the beach, the atmosphere is sexually charged: there are gay couples (the hotel guests) on the sand, sweating, caressing one another. Rosa walks towards the shore as if mesmerised, convinced that her pants are under the sand, or else packed hard with it, "like a piece of torso, a broken statue, the human groin detached, the whole soul gone, only the loins left for kicking by strangers" (48). Stranded on the beach, she can only hear panting noises. The panting comes from the couples on the sand, but Rosa is reliving a traumatic episode, also connected with pants and panting, and with the feeling, which she re-experiences here, of being trapped, abused, treated like an object, like something broken kicked by a stranger that robs you of your soul. Thus, when Persky asks her later on where she had gone, Rosa answers that she had to look for something she had lost, and when further asked by Perky what the lost thing was, she answers: "My life" (55).

Following the pattern of repetitions with slight variations, the scene on the beach takes Rosa back to her rape, but, just as it repeats the rape, it also repeats certain aspects of the episode recounting Magda's death, thus connecting both traumas once again. There is the fence on the beach and in the camp; there are the murmur and voices of the couples on the sand, some of which address Rosa, like the humming voices in the electrified wire of the camp fence. As Magda is thrown onto the fence, the voices in the wire tell Rosa to run towards the child, and she is caught in a double bind, since a part of her wants to run while another part knows she will be shot if she does. It is as if she was suspended between death and the possibility of life. Similarly, on the beach, she absent-mindedly walks near the water and feels the pull of the sea, the pull of an easy death. On both occasions, at the camp and on the beach, the life instinct is stronger, although, as she admits, her life is no life at all. It was cut into two by the war, and that is where she still lives: "Before is a dream. After is a joke. Only during stays. And to call it a life is a lie" (Ozick 2007, 58). A trauma is lasting, it shapes the present and it is, in itself, an open-ended story that shatters chronology. Rosa lives in that frozen time which is the space of trauma.

DEHUMANISED SUBJECTS AND ADDRESSABLE OTHERS

The day after her search for the lost underpants, which ends with her dismissing Persky and calling him thief, Rosa finds the missing underwear curled inside a towel. That same day she decides to have the phone in her room reconnected, which is a symbolic act, a step forwards, towards communication, out of her alienated existence, out of the past. As she tells the receptionist about the phone, the girl hands her a package containing the shawl that she had asked Stella to send. And once again she is pulled by opposing forces, the telephone representing the contrary of what the shawl stands for. Silence and speech are recurrent motifs throughout the text. The shawl stands for silence: it silenced Magda in the camps, ensuring her life, and it silenced Rosa when she stuffed it into her own mouth to avoid screaming with pain at the sight of her daughter's murder. And yet, as Rosa "ingested" the shawl, she was cut off from communication. The silence represented by the shawl has "ingested" her in return. It has become dangerous, maddening, destructive. The shawl has to be removed so that life can go on.

Rosa opens the package containing the shawl that will make her daughter present but, on this occasion, her fetish fails to conjure Magda up as quickly as it used to. Magda *does* eventually come to life, filling the room and appearing to her mother as beautiful as a flower, but then the phone rings and the call significantly brings Rosa back to reality. By the time the receptionist announces Simon Persky the hallucination has come to an end: "Magda was not there. Shy, it ran from Persky. Magda was away" (Ozick

2007, 69). And that is how the story ends. Magda is away, which does not mean she is gone forever, but at least she recedes and Rosa gives a chance to what Persky represents: a present that challenges her obsession with the past. Persky is also that "other" who is willing to listen to Rosa if and whenever she is willing to talk. "Unload on me," he tells her (22). And this is something that nobody had told her before. In order to recompose the shattered pieces that remain of her, Rosa needs someone who does not deny her as a subject, someone who can listen to her: an "addressable other" (Laub 1996, 68). The absence of this empathic listener—and therefore the annihilation of a narrative, the existence of a story that cannot be told, heard, or witnessed—is what constitutes a definitely mortal blow.

Writing on the connection between violence and language in "Deathfugue," Shoshana Felman refers to the violence enacted by the poem in the speech acts of the German commandant, who orders inmates to play music and dig graves. It is already in the very practice of his language that the commandant annihilates the Jews by denying them as subjects (Felman 1996, 31). But this is not the only form of address in the poem. The objectifying, murderous interjections of the German commandant to the inmates meet and clash with what Felman refers to as "the dreaming yearnings of desiring address, the address that institutes the other as a *subject of desire*, and, as such, as a subject for response, of a called for *answer*" (32). Felman is referring here to the direct address to a golden-haired woman (Margareta) and an ashen-haired one (Shulamith), repeated several times throughout "Deathfugue." Significantly, these words are the first the reader encounters as s/he opens *The Shawl*, where the epigraph reads as follows:

> dein goldenes Haar Margarete
> dein aschenes Haar Sulamith
> PAUL CELAN, "Todesfuge"

In Celan's poem, these two lines are placed at the end, separated from the previous lines by a blank. "Deathfugue" thus ends in apostrophe, the closing verses being comparable with the coda that one sometimes finds at the end of a fugue.[7]

Celan chose to bring his poem to a close with the address that institutes the subject and searches for an answer, rather than with the address that objectifies and silences the subject. And yet this ending also brings to the fore the weight of the suffering that will forever prevent a harmonic meeting of Margareta and Shulamith, of the voice that addresses the former and the one that speaks of the latter.[8] The breach is highlighted by the fact that Shulamith's dark hair has turned ashen, which is charged with meanings when seen in connection with the Holocaust and its aftermath. The structural parallelism between the two closing lines somehow underscores an indelible asymmetry. The ending is not redeeming, then, but it at least looks forward, towards a possible opening to a receptive other. "A poem,"

Celan said, "can be a message in a bottle, sent out in the—not always greatly hopeful—belief that somewhere and sometime it could wash on land, on heartland perhaps." In this sense, poems are always on their way toward "something standing open, occupiable, perhaps toward an addressable Thou, toward an addressable reality" (Celan 2001b, 396). The poet inevitably lacks certainty as to whether or not the poem as message in a bottle will wash up on land, but Celan contented himself with the thought that this may happen. *The Shawl* also ends by denying the reader any certainty regarding Rosa's future. She is so deeply wounded that to speak of healing would be to impose on the text our desire for a comforting ending after going through two stories that are distressing and deeply upsetting. And yet, as Rosa opens her door to Persky, the story closes by similarly leaving a door open to possibility: the possibility of a force that will hopefully oppose that of the infernal circles of repetition whose grip the reader is made to feel both in Celan's "Deathfugue" and in Ozick's *The Shawl*. Ultimately, the repetitions in both texts represent the maddening re-enactments that characterise all traumatic experience, thus pointing to repetition as an appropriate vehicle for conveying through literature the oppressive feeling of entrapment that trauma generates, while simultaneously highlighting the need for an addressable other that may counterbalance its deadly effects.

*The research carried out for the writing of this chapter is part of a project financed by the Spanish Ministry of Economy and Competitiveness (MINECO) (code FFI2012–32719). The author is also grateful for the support of the Government of Aragón and the European Social Fund (ESF) (code H05).

NOTES

1. I am using John Felstiner's translation in *Selected Poems and Prose of Paul Celan* (2001, 30–33), where each poem's translation is accompanied by the original in German. The poem is also available on many internet sites. See, for instance, http://www.celan-projekt.de/todesfuge-englisch.html.
2. Adorno's well-known "*Nach Auschwitz ein Gedicht zu schreiben, ist barbarisch*" ["To write poetry after Auschwitz is barbaric" (1981, 34)] was first written in the above-mentioned essay, in 1949. The essay was published singly in 1951, and then collected in *Prismen* (1955).
3. As John Felstiner (1986, 258–59) explains, the "Margarete" of the original German version evokes Heinrich Heine's Lorelei, who combs her golden hair and sings a seductive melody. She is also the Arian ideal of female beauty, the eternal feminine represented by the Gretchen (Margarete) of Goethe's *Faust*, and of Gounod's and Berlioz's. Shulamith (Sulamith in the original) is the comely maiden from the Songs of Songs in the Old Testament, where she is described as a princess who dances on sandaled feet and whose hair is like purple. Felstiner points out that her name may come from a root meaning "whole, complete," akin to *shalom*, "peace," and that she is often seen as a symbol representing the Jewish peoplehood (260).

4. Because the two words sound so similar, there is a sense in which the shawl is also the Shoah, a past injury that is impossible to forget.
5. A "fugue state" would certainly be a good way of describing Rosa's reaction to her rape and her child's murder. According to the *OED*, a fugue can be a flight or escape, and also a musical composition. There is then another meaning, related to the field of psychiatry, where the term "fugue state" refers to a state of mental flight—a flight from one's own identity, one's here and now—as a dissociative reaction to shock or emotional stress. The person's outward behaviour may appear rational while s/he is in that state, but on recovery, memory of the events that provoked the fugue state is repressed.
6. For a wide-ranging analysis of this episode see Levine 2006, 145–54.
7. Various devices may be used to form the conclusion of a fugue. Sometimes the composer may choose to end the fugue with a final flourish, called "code," which is a short section that is added to the piece's major sections and that brings the composition to its close.
8. It is probably the German commandant who addresses golden-haired Margareta (ll. 6, 14, 22, 32), while it must be a different voice that expresses the Jewish yearning for Shulamith (ll. 15, 23). The two addresses, interspersed throughout the poem, are brought together at the end (ll. 35–36), but the voice that utters them is not the same. Felstiner writes these two last lines in German in his translation of the poem.

WORKS CITED

Adorno, Theodor. 1981. *Prisms*. Translated by Samuel and Shierry Weber. Cambridge, MA: MIT Press.
———. 1982. "Commitment." In *The Essential Frankfurt School Reader*, edited by Andrew Arato and Eike Gebhart, 300–18. New York: Continuum.
Celan, Paul. 2001a. "Todesfuge"/ "Deathfugue". In *Selected Poems and Prose of Paul Celan*, edited and translated by John Felstiner, 30–33. New York and London: W.W. Norton.
———. 2001b. "Speech on the Occasion of Receiving the Literature Prize of the Free Hanseatic City of Bremen (1958)." In *Selected Poems and Prose of Paul Celan*, edited and translated by John Felstiner, 395–96. New York and London: W.W. Norton.
Felman, Shoshana. 1992. "Education and Crisis, or the Vicissitudes of Teaching." In *Testimony: Crisis of Witnessing in Literature, Psychoanalysis, and History*. Co-authored by Shoshana Felman and Dori Laub, 1–56. New York and London: Routledge.
Felstiner, John. 1986. "Paul Celan's *Todesfuge*." *Holocaust and Genocide Studies* 1 (2): 249–64.
Harap, Louis. 1987. *In the Mainstream: The Jewish Presence in Twentieth-Century American Literature, 1950s-1980s*. New York: Greenwood.
Horváth, Rita. 1996. "Stella and Rosa." *The AnaChronisT* 2: 257–65.
Klingenstein, Susanne. 1992. "Destructive Intimacy: The Shoah between Mother and Daughter in Fictions by Cynthia Ozick, Norma Rosen and Rebecca Goldstein." *Literature* 11 (2): 162–73.
Lang, Berel. 2000. *Holocaust Representation: Art within the Limits of History and Ethics*, Baltimore and London: Johns Hopkins University Press.
Laub, Dori. 1992. "Bearing Witness, or the Vicissitudes of Listening." In *Testimony: Crisis of Witnessing in Literature, Psychoanalysis, and History*.

Co-authored by Shoshana Felman and Dori Laub, 57–74. New York and London: Routledge.

Levi, Primo. 1989. *The Drowned and the Saved*. Translated by Raymond Rosenthal. New York: Vintage.

Levine, Michael. 2006. *The Belated Witness: Literature, Testimony, and the Question of Holocaust Survival*. Stanford, CA: Stanford University Press.

New World Encyclopedia contributors. 2008. "Fugue." *New World Encyclopedia*. Accessed March 10 2011. http://www.newworldencyclopedia.org/p/index.php?title=Fugue&oldid=778598.

Ozick, Cynthia. 2007. *The Shawl*. London: Phoenix.

12 Of Ramps and Selections
The Persistence of Trauma in Julian Barnes' *A History of the World in 10 ½ Chapters*

Jean-Michel Ganteau

A History of the World in 10½ Chapters is certainly, with *Flaubert's Parrot*, the best-known and most glossed upon of Julian Barnes' novels. When released, it was hailed as an irresistibly witty piece of historiographic metafiction, and most reviewers and critics commented on its hybrid nature. It was essentially valued for its mixture of genres and modes, its toying with the boundary between fiction and history, its narrative instability, in other words its frankly disruptive, metafictional inspiration germane to the formalist view of postmodernism so influential throughout the 1990s (Guignery 2006, 61–62). At the time, most of the criticism concentrated on the text's ambiguous status: was it a novel, or did it rather lean towards a collection of stories? Such were matters that critics like Moseley and Guignery among many others engaged with (Moseley 1997, 110; Guignery 2001, 28). Yet, behind—or beyond—the aesthetic considerations devoted to the hybrid form and the ludic dimension of the narrative, most commentators did underline the seriousness of purpose and, more specifically, the *gravitas* in which the whole of the project was steeped. Moseley made a point about the corrective drive of the novel, underlining that Barnes' project was to problematise the traditional vision of history as that of winners (Moseley 1997, 120), Rubinson promoted the vision of *A History of the World in 10 ½ Chapters* as narration of an alternative understanding of history (Rubinson 2000, 163), while Buxton read it in the light of Benjamin's conception of history, suggesting that Barnes "prefer[red] an apocalyptic philosophy of history rooted in a vehement disavowal of historical progress" (Buxton 2000, 58).

Such critical orientation may have been partly fostered by the influence of what has been termed "the ethical turn" in literary production and criticism over the last two decades, for in fact, at the root of such explorations lies the idea that the experimental drive that informs the narrative is meant as the privileged expression of an ethical concern with the limits of conventional historiography, and that a revisionist mode predicated on a "decentering of perspective" (Guignery 2006, 69–70) might well provide the fictional presentation of an ethics of truths and, more generally, of an ethics of alterity taking into consideration not only the history of the victims—as opposed to that of victors—but also *the other* of conventional history. One step further, the catastrophic dimension of history was underlined in most analyses when accounting for the various threads or motifs linking the apparently disconnected chapters, among which prominently figure voyages and maritime catastrophes (Guignery 2001, 91), but also, most obviously, the ark motif (complete with the *topoi* of embarkation and disembarkation). This recurrent image goes hand in hand with that of the separation of the clean from the unclean, as analysed by Moseley among many others in terms of "sorting out" the normal from the abnormal (Moseley 1997, 117–18), a figure referred to in terms of "selective disembarkation" by Buxton (Buxton 2000, 71) who considers the motif of the voyage as catastrophe one of the most tightly federating elements in the narrative: "the human voyage is an unpiloted drift from disaster to disaster, where the innocent perish and the survivors crew a ship of fools. History is merely one retreat from catastrophe that blunders into another" (Buxton 2000, 61). This is proof enough that the criticism on *A History of the World in 10½ Chapters* has not confined itself to the analysis of the *modes* of revisionism or the *process* of ironisation as cut out from a context, but has been very much aware of the *contents* of the historical material at hand. Seen in this light, it is a piece of literature committed to testimony.

However, very few commentators have applied trauma theory to the novel, hardly any in fact but Onega who devoted an article to what she called "the nightmare of history" in *A History of the World in 10½ Chapters*. She addressed the issue of traumatised narrators finding indirect means and new forms to convey their meanings, thus effectively positing an ethics of experimentation, and emphasising the primacy of experimentation when it came to expressing the experience of trauma (2008, 356, 359). Taking my lead from her work, I will argue that *A History of the World in 10½ Chapters* is essentially a trauma narrative, not so much as regards individual trauma (which is admittedly staged in Chapter 4, "The Survivor", through the character called Kathleen Ferris) but rather as a collective, historical trauma, i.e. the trauma of shipwreck, of what may be seen as the impending nuclear holocaust, of religious fundamentalism, and essentially—though indirectly—the trauma of the Shoah. I will further argue that, among the various apocalyptic evocations in play, the persistence of images and after-effects of the Shoah is what characterises the narrative, a motif that has been spotted by commentators and that is perceptible in most chapters, through the figure of selection that federates constellations of related images of survival and extermination. I intend to show that what the narrative performs is the persistence of trauma, which is perceptible through a highly repetitive, fragmented and echoic text.

In an interview with Guignery, Barnes addresses the question of the evocation of the Shoah in relation with one of the sections in the "Three Simple Stories" chapter, the one evoking the tragic cruise of the Saint Louis, a pleasure ship with several hundred Jews on board that sailed to Havana and back to Europe in the spring of 1939, a few weeks before the beginning of WWII, taking back part of its cargo of undesirable emigrants to Germany—hence to inescapable extermination. This episode he evokes in the following words:

> I think I would have also felt, if I had fictionalised it, that I was unfairly playing with people's lives. I think when you get near areas like the Holocaust, unless you are particularly a special witness of them, like, say, Primo Levi, then I think you have to be very, very careful about using them in any way. They're almost sacred subjects. (Guignery and Roberts 2009, 58)

Now, even if one should beware of authors' pronouncements, I think that such a statement gives a clue as to the way in which the issue is tackled circuitously and indirectly in *A History of the World in 10½ Chapters*. True, the topic is explicitly alluded to in some passages, as in Chapter 2, providing a fictional rendition of the Achille Lauro crisis (Barnes 1990, 55–56), but nowhere is it more explicitly tackled than in Chapter 5, "Shipwreck", which documents and fictionalises the disastrous journey that provided the material for Géricault's *Raft of the Medusa*. When dealing with the master's representation of the survivors, the narrator wonders why they look so healthy after the atrocious hardships they had to go through, and provides a tell-tale analogy:

> When television companies make drama-docs about concentration camps, the eye—ignorant or informed—is always drawn to those pyjamaed extras. Their heads may be shaven, their shoulders hunched, all nail varnish removed, yet still they throb with vigour. As we watch them queue on screen for a bowl of gruel into which the camp guard contemptuously spits, we imagine them offscreen gorging themselves in the catering van. (Barnes 1990, 136)

Such words obviously echo those of Barnes in interview and re-formulate the ban on the poetic and fictional representation of trauma voiced by, among others, Theodor Adorno and Jacques Lanzmann. They also connect with certain allusions that might pass unnoticed, like the reference to the "Mussulman" (in the phrase "Christian and Mussulman") that, in the context of a chapter offering a pastiche of nineteenth-century English, might pass as archaism (Barnes 1990, 155). Still, the term is highly evocative of that typical figure of the death camp analysed among others by Primo Levi and Giorgio Agamben as the Musselman, a word referring to the inmate who is at the end of his/her tether and has entered the zone from which there will be no return. The Musselman stands as the emblem of the very temporary survivor, whose fate is sealed, and who will not survive beyond a few days or weeks.

Such allusions, one realises, concern less the universe of the concentration camp than the ecology of the death camp, to take up Robert Jay
Lifton's metaphor (Lifton 1986, 27 *et passim*), and they do pave the interpretative way towards the partially hidden ethical agenda of the novel,
figuring some sort of textual unconscious, which is the evocation of the
trauma of the Shoah. Such a dimension is partially hidden but flaunted
by the means of a historical allegory predicated on the parodic revision
of the Flood passage in Genesis, making the novel a fiction of the archive.
From this perspective, the central figure of Noah and his Ark becomes an
image to which the first chapter is devoted and which radiates as a motif
throughout the other 9 ½ chapters.

It is well-known that the opening of *A History of the World in 10½
Chapters* is, in Bakhtinian terms, a piece of *parodia sacra*, devoted to the
debunking of a sacred text. One remembers that it is narrated from the
point of view of a stowaway, a woodworm, that embarked unbeknownst
to Noah and his family, and whose voice hyperbolically represents that
of the minor participants on the stage of history. In full revisionist swing,
the humble narrator draws the portrait of Noah as cantankerous, cruel,
totalitarian patriarch and presents the picture of the ark as a prison ship in
which inspections and punishments are perpetrated on a purely gratuitous
basis (Barnes 1990, 6, 7, 8, 9, *et passim*), and whose essence resides in the
recurrence of selections. In fact, the world of the ark (before embarking
and throughout the voyages) is one of selections of an even starker type
than natural selection, as the best of animals have to be selected to access
the ship, whose gangways are evoked in terms of "ramps" (Barnes 1990,
26)—a word reminiscent of another locus of selection, i.e. Auschwitz, as
indicated by Lifton among others (Lifton 1986, 15 *et passim*). This is of
course highly ironical as the ark is originally a salvaging symbol, whereas
in Barnes' novel it is turned into a figure of destruction as, instead of healing and uniting, it cleaves and separates. However, what the ark motif aptly
allows for is the evocation of the collectiveness of the experience, as the
representatives of the animal kingdom are allowed on board as emblematic
of all species. The synecdochic vein means that the disappearance of a pair
stands for that of a whole species or group—an efficient way to predicate
collective trauma on individual trauma, allowed by the selection trope—so
much so that the sorting out of the clean from the unclean that becomes
one of the main motifs of the first chapter and of the book as a whole quali-
fies as allegory, in which the selection of a pair/group to be saved or to be
exterminated indirectly evokes the issue of genocide. To take up Onega's
words, through the allegory "the unbroachable motifs of racial hatred and
genocide are made representable" (2008, 363). In the light of such textual
evidence, the selection and extermination of the crossbreeds (Barnes 1990,
26), together with the killing of the sick (13, 26)—a motif that will recur in
the "Shipwreck" chapter, when the survivors on the raft will have to make
the decision to execute the weaker of them so as to consume their flesh

(121, 128)—becomes specifically redolent of Nazi eugenics and of the bio-medical vision that Lifton sees as being essential to Nazism in the form of applied biology (Lifton 1986, 157). This is corroborated by the evocation of insecticide or vermicide (one remembers that the narrator of Chapter 1 is a woodworm) in terms that call forth the spectral presence of the gas chamber, complete with pellets of Ziklon B, or the medical experiments perpetrated by medical units: "long before the days of the fine syringe filled with a solution of carbolic acid in alcohol, long before creosote and metallic naphthenates and pentachlorphenol and benzene and para-dichlor-benzene and ortho-di-chloro-benzene" (Barnes 1990, 18).[1]

This vision seems to come to a head as the text becomes concerned with the consequences of such acts, comparing the sacred ark to a death camp and envisaging the salvaging of the human race and of the animal world within a context of historical bio-medical purification or cleansing. This leads in fact to a despondent remark on the big holes in the animal kingdom, in which one may see a reference to the missing families and huge gaps in the demography and genealogies of European Jews (and other victims of the final solution, like gypsies, homosexuals, the mentally and physically handicapped, among others) after World War Two, what the stowaway narrator calls "leaps in the spectrum of creation" (Barnes 1990, 13). Elsewhere the consequence that is alluded to is that of "the traumatisation of entire species" (22), a description that cannot but recall the issue of collective trauma—more specifically the collective trauma of a persecuted, martyred community. Elsewhere, Noah's selection is compared to the blindly rigorous laws of natural selection in explicit terms: "What do you call that—natural selection? I'd call it professional incompetence" (7). Such analogy offers a striking parallel between the various selection processes at work in the narrative and in so doing helps orient the satire towards the principles and practice of Nazi selection, denouncing its gratuitous blindness, its total dehumanisation and its elemental sense of overreaching. And perhaps one could go so far as to claim that the parallelism between Barnes' and Lifton's books goes so far as to envisage the problem of perpetrators' trauma—one of the main lines of investigation in *The Nazi Doctors*—when Noah's alcoholism is described as evasion from responsibility, in the wake of the numbing and doubling protective devices that Lifton sees as characteristic of the Nazi doctors in charge of selection and other gruesome tasks (Lifton 1986, 9, *et passim*): "Old Noah had always enjoyed a few horns of fermented liquor in the days before Embarkation: who didn't? But it was the Voyage that turned him into a soak. He just couldn't handle the responsibility" (Barnes 1990, 29–30).

I would then argue that the concentration of selection-related motifs, and its radiation throughout the book performs a repetition surfacing in heterogeneous contexts around the central figure of the survivor. Such obsessive recurrence is notoriously one of the characteristics of trauma (Freud 2001, 18 *et passim*; Caruth 1996, 2) and of trauma fiction as analysed by Anne

Whitehead among others (Whitehead 2004, 85).[2] From this point of view, *A History of the World in 10½ Chapters* may be considered a particularly successful example of trauma narrative and traumatic realism, in Rothberg's acceptation of the term, i.e. a mode of narration that relies on experimental (or at least subversive) devices the better to express the extremity of traumatic experience. Barnes' text, I would contend, provides the reader with an atypical instance of traumatic realism in that it manages, circuitously and discretely, to evoke the paroxysmal symptoms of trauma, a state characterised by the paradox according to which the more violent the trauma, the more difficult the access to its origins (Caruth 1996, 92). In other words, I would claim that what Barnes succeeds in doing is precisely presenting (short of representing) the hesitation according to which trauma fails to know itself and represent itself (Amfreville 2009, 44; Press 1999, 52–54). In so doing, he banks on a new form of referentiality turning its back on more traditional mimetic devices so as to perform what critics like Rothberg (2000) and Luckhurst (2008) have defined as "traumatic realism."

By resorting to the indirection of diluted allegory, this is what *A History of the World in 10½ Chapters* precisely achieves, i.e. it evokes the latency of a traumatic event whose effects are still to be felt, manifesting themselves repeatedly, as barely emerging ripples, thus being iconic of the symptoms of trauma, as if, in Wallhead and Kohlke's terms, trauma were displaced to the textual unconscious (2010, 232). Imitation of effect (as distinct from imitation of aspect) would thus be the performative basis on which the event of the Shoah is metonymically evoked, i.e. through the resort to that trope, the metalepsis (one of the many versions of metonymy), which calls forth an event either through its causes or its consequences, the second option being the one privileged here. In so doing, Barnes re-negotiates the conventions of realism by turning his back on conventional representation, preferring presentation or performance, which carries the intensifying characteristics that Whitehead sees at work in trauma fiction (2004, 86) and helps mix the referential with the referentially disruptive (Rothberg 2000, 3–5). Through the rehearsal of such motifs, what is suggested is a sense of haunting, in conformity with Freud's vision of trauma as the belated manifestation of an event that cannot be remembered and has to be repeated in the present (Freud 2001, 18). Such ceaseless return of the same in the garb of the other or in heterogeneous contexts achieves a purely uncanny effect and, in so doing, warrants the advent of an ethical reading that Rothberg sees as the hallmark of traumatic realism, as the readers "are forced to acknowledge their relationship to posttraumatic culture" (2000, 103). One step further, this may account for the highly fragmented form of the narrative, and its lack of narrative cohesion (one remembers the central hesitation as to its status, and its being considered as a collection of short stories as much as a novel), together with its rejection of linearity which are characteristic of trauma fiction and testify to trauma's aversion to chronology. *A History of the World in 10½ Chapters* presents the reader with unhinged, traumatic time, in which belatedness is fully at work, in the original meaning of the term, i.e. not only as postponement, but also as the modification of the first occurrence by the later occurrences determined by the compulsion to repeat (Mieli 2001, 44–45; Amfreville 2009, 75).[3]

For in fact, the chronology of *A History of the World in 10 ½ Chapters* appears to be in crisis, as if under the sway of an echoic structure that seems to have gone awry, since not only does the Shoah surface in narratives dealing with contemporary events (those in Chapters 2, 4, 7, 8), but it also looms over anterior periods of history (as presented in Chapters 1, 3 and 5) that it revisits in terms of *Nachträglichkeit*, as if the permanence or persistence of trauma were such that it spread in all directions and became both ubiquitous and a-temporal, the whole of history as traumatic history becoming always already post-traumatic. Such seems to be the main effect of a narrative that looks as if it were going in concentric circles, like Katherine Ferris in the "Survivor" chapter, alone with her cats on her boat, circumnavigating the port without setting off towards the open sea. The impression that the reader is left with is that of a narrative in which traditional temporality does not apply, a narrative governed by the law of repetition of some dead time that does not elapse any more, but is stuck in a *Zwischen* or an "in between" that characterises the temporality of trauma. In Davoine and Gaudillière's terms, there is a suspension of the flow of time and time stops, or rather the reader is confronted with a time that does not elapse (Davoine and Gaudillière 2004, 18). The time of *A History of the World in 10½ Chapters* is both traumatised and traumatic time in that it only exists in relation to an original effraction (Gibson 1999, 182). It is a time that stops and starts radiating in all directions, to suggest that the catastrophes evoked in the text, and more particularly the Shoah, are not inscribed in history (Davoine and Gaudillière 2004, 126) or rather that their inscription in history remains unstable. Such a time, so warped and full of itself, would thus appear as the main symptom of the inassimilability of the traumatic effraction, the big caesura that rifts the twentieth century and world history as a whole. I would thus claim that Barnes' narrative performs the impossibility of moving beyond trauma and metonymically produces the portrayal of a posttraumatic age in which every textual event is a symptom. Less than a subversive version of history, it offers a traumatised and traumatic vision of history, in keeping with Barnes' revisiting of Marx's famous statement: "I think history does repeat itself, but never according to such simple rules, first time as tragedy, second time as farce. Mostly it repeats itself first time as tragedy and second time as tragedy, or, alternatively, first time as farce and second time as farce" (Guignery and Roberts 2009, 57). With the sole reservation that *A History of the World in 10½ Chapters*, as trauma narrative, *rehearses* tragedy, i.e. repeats it looking both backwards and forwards to the next occurrence of the repetition, adopting the movement of the gyre or spiral, as if trauma were always beyond itself, as if there were indeed very little possibility of moving beyond trauma.

And still . . . The narrative is dominated by an inescapable energy that counteracts the effects of traumatic entropy and hosts passages whose levity of tone cannot be denied, as if the traumatic model evoked above were but partially valid. In fact, there are intimations of a way of moving beyond trauma. Said differently, *A History of the World in 10 ½ Chapters* is also concerned with healing and coming to terms, in more ways than just one and, in Caruth's terms, the narrative of trauma reveals as much as it hides, so much so that the text figures forth what it cannot think out (see her chapter in the present volume). Figuration would thus be the way allowing for the move beyond trauma.

At the most basic level, it may be said that Barnes' narrative does provides an instance of traumatic realism that imitates (and perhaps more fundamentally mimics) the coming to consciousness of the traumatic catastrophe. And I would argue that this is a way of indicating the move from a phase of mere acting out to one of possible working through, while narrativisation in itself favours and allows for the transition. Seen in this light, all the textual symptoms analysed above would point to some degree of revelation and acceptance of the traumatic event that ceases to be denied and silenced and that comes to consciousness in a tentative way. *A History of the World in 10 ½ Chapters* would thus present the reader with a stammering, incipient phase of working through, in which time would be not so much fixed and motionless as on its way towards integration, and in which latency would be trembling on the brink of inchoation.

In other words, the text would work less on the model of incorporation than according to the rules of introjection, to take up an opposition lying at the core of Abraham and Torok's work. The narrative that the reader is presented with is in the process of coming to terms with the lies and silences of history and is certainly not content to accept them, rehearsing them and incorporating them. If the individual model posited by Abraham and Torok (2011, 227–37) may be applied to that of history, it may be said that *A History of the World in 10 ½ Chapters* refuses to remain the tomb or crypt of the lying biblical and historical texts. Just as the incorporation of an object as compensation for the loss of a loved one may transform an individual into a tomb in some extreme pathological cases, so some narratives might be considered crypts of other narratives, whether fictional or not. This implies that, in many respects, the intertextual or hypertextual revisiting imitates a compulsion to repeat, when the repeating narrative gives no clue as to its being conscious of the repetition. In such cases of direct and massive influence, the figure that comes to mind is that of ventriloquism, which is compatible with instances of pastiche used without distance. In such cases, the contemporary text could be said to be the crypt in which the previous one lies incorporated, the second text being mere rehearsal, itself being deprived of a textual unconscious. But things look different with *A History of the World in 10 ½ Chapters*, in which the use of parody signposting itself as such introduces a great deal of distance and awareness of the repetitive process, so much so that it can be said to perform a move both beyond pure denial and beyond pure acting out. Barnes' narrative is not mere ventriloquism and possession but repetition with a critical, ethical distance and it points at its own work as incipient working through. In other words, instead of merely incorporating otherness so as to lose its own singularity and become pure symptom (Abraham and Torok 2011, 118), it seems to favour a more introjective model in which the other text is taken into account (almost taken on board, if the maritime metaphor may be extended here) under the guise of an opening to otherness and a widening that eschews strict incorporation and mere repetition.

Similarly, it may be said that *A History of the World in 10 ½ Chapters* refuses to be haunted by a phantom, in Abraham's acceptation of the term (Abraham and Torok 2011, 426–33). What I mean to suggest is that, by transposing a clinical model to a literary context, in conformity with Abraham's considerations on the ghost, a text may be haunted by a silence that was at work in previous texts or, to extend the anthropomorphic analogy, in texts of a previous generation. Of course, the Shoah was thematised in various documentary, testimonial texts in the years after the Second World War, but not overwhelmingly so, and most of the fictional production, in Britain and in the Western world at large, was fairly discreet about the matter till the mid-1980s. My point is that an early piece of historiographic metafiction like *A History of the World in 10 ½ Chapters* refuses such silence and makes it visible or vocal by resorting to the textual symptoms evoked above, and through the multiplication of metafictional allusions to previous narratives that thematise the notion of selection and multiply references to its coming to consciousness. From this point of view, it may be argued that the narrative refuses to be haunted by the ghost of previous Shoah inspired (fictional) testimonies and refuses to be dominated by the stranger within (Abraham and Torok 2011, 448). Thanks to the distance inherent in the parodic and metafictional reworking of previous narratives of selection, what emerges is a commitment to voicing the shameful, buried secret so as to exorcise it (Abraham and Torok 2011, 448). More specifically here, I would say that the reworking of a fragment of Genesis as paradigmatic selection narrative is a way to summon what might appear to be a phantom text, in Royle's terms, i.e. a text that is neither present nor absent, and a text which perhaps cannot be singled out as intertext or hypotext, but that emerges in between texts so as to counteract the power of silence and lay the phantom born out of the previous generation's silence (Royle 2003, 277–81). Here, the phantom text of the Shoah as selection becomes an arch-representative of fictional testimony in which a form of nescience or living-dead science both forbids knowledge and triggers off a quest for knowledge (Abraham and Torok 2011, 448–49), with the reservation that in this case, the possibility of knowledge is privileged to such an extent that the quest is permanently on the brink of coming to completion. The story that Barnes' narrative tells is that of the laying of the phantom, and it performs such laying even while it tells its story.

Ultimately, what seems to be at the heart of *A History of the World in 10 ½ Chapters* is the affirmation, all discretion jettisoned, of a sense of expectancy, in Davoine and Gaudillière's acceptation of the term, i.e. "the actual hope for life when life has been banished from the horizon. [The term 'expectancy'] sketches the outlines of an otherness that awaits you against all expectation, all logic, all common sense" (Davoine and Gaudillière 2004, 209). More precisely, what is at the heart of such expectancy is that there is someone else, some other bearing witness to the idea of a fundamental solidarity that the two French psychoanalysts couch in the following terms:

> Expectancy is the expectation that an other will take over for you when you are exhausted [. . .]. The water or potato peelings that save your life, the eyes that cry out to throw yourself on the ground when danger arises, the warmth of a back when, in normal times, the lack of privacy would be unbearable: all survivors bear witness to the impossibility of surviving alone (Delbo 1970a, b), without another, present or, if necessary, hallucinated. (Davoine and Gaudillière 2004, 210)

I would argue that, ultimately, this is what *A History of the World in 10 ½ Chapters* comes up with as a solution to move beyond individual trauma and the trauma of history in the famous half chapter entitled "Parenthesis", a generically unstable chapter voiced by a narrator who refers to himself as Julian Barnes, whose theme is love and connubial bliss in faithfulness, but which prefers the medium of what it calls "love prose" (Barnes 1990, 228) to the more hackneyed conventions of sentimental romance. In this surprising, unclassifiable half chapter Barnes takes the risk of using the vulnerable form of his specifically coined and honed out "love prose" (unmasked first person, claim to sentimentality and what may be read as mawkishness, revisiting of one of the most trite themes in the literary tradition) so as to warn the reader against the dangers of false history and false love (239–40). He reminds the reader that "the heart is not heart shaped" (232) and that bad history is but "soothing fabulation" (242), while on the contrary there is such a thing as truthful history (240). One step further, he claims that what characterises love is not only its uncompromising need for and revelation of truth (as encapsulated in the oxymoronic phrase "*Lying in bed, we tell the truth*" [240]), love being the only value to which we must commit ourselves and be faithful. The half chapter, so conspicuous on account of its hybrid, heterogeneous structure and enigmatic titular inscription, ends up with an ethical call for faithfulness to human love (246), a call that risks naivety and mawkishness and takes pride in the vulnerability of its form so as to forcefully make a point, suggesting a solution for the contemporary subject's coming to terms with the nightmare and trauma of history.

Now, such a solution might sound diminutive and ill adapted, consigned to the individual sphere, an instance of the "think small" type of attitude. However, Barnes insists that his is a real solution as opposed to the mere "soothing fabulation" of history making (Barnes 1990, 242). Radically faithful, furiously ethical love as conceived of by the narrator and author is not a matter of mere soothing but of healing, because of its capacity for "imaginative sympathy" (243) that makes the individual sensible and open to—in other terms vulnerable to—the face of the other and the traumas of the other. And Barnes addresses the sublime, traumatic welter of history with one solution, i.e. love as responsibility, including political and historical responsibility:

> What else will love do? If we're selling it, we'd better point out that it's a starting point for civic virtue. You can't love someone without imaginative sympathy, without beginning to see the world from another point of view. You can't be a good lover, a good artist or a good politician without this capacity (you can get away with it but it's not what I mean). Show me the tyrants who have been great lovers. (243)

Even while it mimics trauma, thus making visible and audible the silences of history and more especially imitating the coming to consciousness of one of the great silences of history, *A History of the World in 10 ½ Chapters* is committed to the presentation of a phase of historical trauma hesitating on the verge of acting out and working through. From this point of view it describes trauma on the cusp, and does so through a poetics of the cusp, using a hybrid, vulnerable form to present trauma as incipient healing, pointing to a glimmer of hope, in the direction of some state beyond trauma.

NOTES

1. From this point of view, *A History of the World in 10½ Chapters* might be considered a companion piece to Martin Amis' *Time's Arrow*, with its evocation of life in Auschwitz and of the role of the doctor in an Auschwitz world full of selections, eugenics, medical experiments, and chemical substances. For a thoroughly convincing evocation of this novel in terms of trauma criticism and theory, see María Jesús Martínez Alfaro's "Where Madness Lies: Representation and the Ethics of Form in Martin Amis' *Time's Arrow*" (Martínez Alfaro 2011). For an analysis of the intertextual relationships between *Time's Arrow* and *The Nazi Doctors*, see Ganteau 2010.
2. One may remember that Whitehead singles out intensification, intertextuality, and repetitions as the main formal traits of trauma fiction—by which she does not mean traumatic realism, as she makes clear on page 84 (2004, 84–87).
3. For a detailed, fully convincing account of the various translations and meanings of Freud's *Nachträglichkeit* and for a study of its effects, see Laplanche, more specifically his emphasis on the dual movement of *Nachträglichkeit* as both the presentification of the past and the subject's return to the past

(2006, 26–27) ; and his vision of the warped time of trauma as spiral as opposed to arrow (89).

WORKS CITED

Abraham, Nicolas, and Maria Torok. (1987) 2011. *L'écorce et le noyau.* Paris: Flammarion.

Amfreville, Marc. 2009. *Ecrits en souffrance.* Paris: Houdiard.

Barnes, Julian. (1989) 1990. *A History of the World in 10 ½ Chapters.* London: Picador.

Buxton, Jackie. 2000. "Julian Barnes's Theses on History (in 10 ½ Chapters)." *Contemporary Literature* 14 (1): 56–86.

Caruth, Cathy. 1996. *Unclaimed Experience: Trauma, Narrative and History.* Baltimore and London: The Johns Hopkins University Press.

Davoine, Françoise and Jean-Max Gaudillière. 2004. *History Beyond Trauma.* Trans. Susan Fairfield. New York: Other Press.

Freud, Sigmund. (1920) 2001. "Beyond the Pleasure Principle." In *The Standard Edition of the Complete Psychological Works of Sigmund Freud*, XVIII. Edited by James Strachey, 7–64. London: Vintage.

Ganteau, Jean-Michel. 2010. "De l'allusion au commentaire: le travail de la citation (*Time's Arrow* et *The Nazi Doctors*)." In *L'intertextualité dans le roman contemporain de langue anglaise.* Edited by Jocelyn Dupont and Emilie Walezak, 123–37. Perpignan: Presses de l'Université de Perpignan.

Gibson, Andrew. 1999. "Crossing the Present: Narrative, Alterity and Gender in Postmodern Fiction." In *Literature and the Contemporary: Fictions and Theories of the Present.* Edited by Roger Luckhurst and Peter Marks, 179–86. Harlow: Longman.

Guignery, Vanessa. 2001. *Julian Barnes: L'art du mélange.* Pessac: Presses Universitaires de Bordeaux.

———. 2006. *The Fiction of Julian Barnes: A Reader's Guide to Essential Criticism.* Houndmills, Basingstoke: Palgrave-Macmillan.

Guignery, Vanessa, and Ryan Roberts, eds. 2009. *Conversations with Julian Barnes.* Jackson: University Press of Mississippi.

Laplanche, Jean. 2006. *L'Après-coup. Problématiques VI.* Paris: Presses Universitaires de France.

Lifton, Robert Jay. 1986. *The Nazi Doctors: Medical Killing and the Psychology of Genocide.* New York: Basic Books.

Luckhurst, Roger. 2008. *The Trauma Question.* London and New York: Routledge.

Martínez Alfaro, María Jesús. 2011. "Where Madness Lies: Representation and the Ethics of Form in Martin Amis' *Time's Arrow*." In *Trauma and Ethics in Contemporary British Fiction.* Edited by Susana Onega and Jean-Michel Ganteau, 129–56. Amsterdam: Rodopi.

Mieli, Paola. 2001. "Les temps du traumatisme." In *Actualité de l'hystérie.* Edited by André Michels, 43–60. Toulouse: eres.

Moseley, Merritt. 1997. *Understanding Julian Barnes.* Columbia: University of South Carolina Press.

Onega, Susana. 2008. "The Nightmare of History, the Value of Art and the Ethics of Love in Julian Barnes's *A History of the World in 10 ½ Chapters*." In *Ethics in Culture: The Dissemination of Values through Literature and Other Media.* Edited by Astrid Erll, Herbert Grabes and Ansgar Nünning, 355–68. Berlin and New York: de Gruyter.

Press, Jacques. 1999 *La perle et le grain de sable: Traumatisme et fonctionnement mental*. Lausanne: Delachaux et Niestlé.

Rothberg, Michael. 2000. *Traumatic Realism: The Demands of Holocaust Representation*. Minneapolis and London: University of Minnesota Press.

Royle, Nicholas. 2003. *The Uncanny*. London and New York: Routledge.

Rubinson, Gregory T. 2000. "Julian Barnes's 'A History of the World in 10 ½ Chapters'." *Modern Language Studies* 30 (2): 159–79.

Wallhead, Celia, and Marie-Luise Kohlke. 2010. "The Neo-Victorian Frame of Mitchell's *Cloud Atlas*: Temporal and Traumatic Reverberations." In *Neo-Victorian Tropes of Trauma: The Politics of Bearing After-witness to Nineteenth-century Suffering*. Edited by Marie-Luise Kohlke and Christian Gutleben, 217–52. Amsterdam: Rodopi.

Whitehead, Anne. 2004. *Trauma Fiction*. Edinburgh: Edinburgh University Press.

13 The Trauma of Anthropocentrism and the Reconnection of Self and World in J. M. Coetzee's *Dusklands**

Susana Onega

EXPERIMENTALISM AND THE UNUTTERABILITY OF TRAUMA

J. M. Coetzee earned a reputation as an experimental writer already with his first novel, *Dusklands*. Published in 1974, at a time when South African literature was expected to be realistic and politically engaged, *Dusklands* baffled the readers' expectations with its stylistic self-reflexivity, thematic dispersion and structural fragmentation. Dominic Head summarised the general outlook on it in 1997 when he wrote that, by the time of publication of his study, "[i]t ha[d] become a truism in criticism of Coetzee that *Dusklands* (1974) introduces a new postmodernist strain in the novel from South Africa" (28). However, though general, this view was not unanimous. Two years earlier, Ronald Granofsky had expressed his reluctance to apply the label of postmodernism to Coetzee, preferring to place him (together with Thomas Pynchon)in an intermediary position between modernism and postmodernism.[1] More recently, Derek Attridge has argued that "Coetzee's work follows on from Kafka and Beckett, not Pynchon and Barth," so that "[i]t would be more accurate [. . .] to characterize it as an instance of 'late modernism,' or perhaps 'neomodernism'" (2004, 2). And he has responded to adverse criticism on Coetzee's work with the argument that modernist self-reflexivity is the ethical expression of a desire to break up the constraints imposed on transparent literature by the dominant discourse, and to give voice to what is excluded by this discourse. To complicate the issue further, yet another critic, Anne Haeming, has recently highlighted Coetzee's constant recourse to textual genres such as autobiographies, diaries, chronicles and letters, and his need to "draw attention to the *edges of texts* and, consequently, *the edges of fact and fiction*," in what she describes as a "compulsive search for authenticity" and an overriding desire "to communicate truthfulness" (2009, 174). Together with self-reflexivity and fragmentariness, Coetzee's novels also display, then, a contrary impulse to recreate the realism-enhancing mechanisms employed by the novel at the time of its birth. This generic hybridity may be interpreted as proof of the writer's efforts to create his own novelistic form, one capable of demystifying received notions of reality and truth and of foregrounding the ideologically-charged nature of human discourses, in particular, the discourse of modern history, which, echoing Theodor Adorno, Coetzee describes in *Doubling the Point* as "the spirit of the barbarian [...], which is pretty much the same thing as history-the-unrepresentable" (1992, 67).

In an interview with David Attwell, Coetzee casts significant light on his own outlook on history when he comments on the attitudes of Milan Kundera, James Joyce and Zbigniew Herbert. He admits that a part of him longs to be as "socially irresponsible" as Kundera was when he said in the Jerusalem Prize lecture in 1985 that: "Today, when politics have become a religion, I see the novel as one of the last forms of atheism" (1992, 68, 67), but then adds that he feels incapable of following Kundera's example, just as he has never known "how seriously to take Joyce's—or Stephen Dedalus'—'History is a nightmare from which I am trying to awake'" (67). To Coetzee, then, a refusal of commitment is a non-issue. What is more, Coetzee thinks that there are responses to history that fail to meet ethical standards. Commenting on Zbigniew Herbert's poem "Five Men,"; where—presumably in opposition to Adorno's famous dictum: "To write poetry after Auschwitz is barbaric" (1981, 34)—the Polish poet argues for the power of poetry to oppose Stalinism, Coetzee reflects:

> In Poland one can still hold such beliefs; and who after the events of 1989, would dare to scorn their power? But in Africa [...] the only address one can imagine is a brutally direct one, a sort of pure, unmediated representation; what short-circuits the imagination, what forces one's face into the thing itself, is what I am here calling history. "The only address one can imagine"—an admission of defeat. *Therefore*, the task becomes imagining this unimaginable, imagining a form of address that permits the play of *writing* to start taking place. (1992, 67–68)

As this quotation makes clear, the form of Coetzee's works is dictated by the struggle between an (evidently Beckettian) admission of failure counteracted by the ethical obligation to give voice to the unimaginable and unutterable traumas of modern history.

In a book entitled *Traumatic Realism: The Demands of Holocaust Representation*, Michael Rothberg (2000) explains how the perplexities of understanding and representing the Holocaust, with its peculiar combination of ordinary and extreme elements, has led to three fundamental demands: "a demand for documentation, a demand for reflection on the formal limits of representation, and a demand for the risky public circulation of discourses on the events" (7). These demands have in turn triggered off the need to rethink "the categories of realism, modernism, and postmodernism [...] not only as styles and periods [... but also] as persistent responses to the demands of history" (9). His contention is that the overlapping of representational modes constitutes a "complex system of understanding" aimed at forcing readers to think history in relational terms and so perceive the relationship between (realist) documentation, (modernist) self-reflexive aesthetic form, and (postmodernist) public circulation. Roth-berg gives as a visual example of the tension created by the conjunction of these three modes a drawing by Art Spiegelman representing a dead mouse, Maus and Micky Mouse that first appeared in *Tikkun* magazine, accompa-nying the article "Saying Goodbye to Maus" (xiii). And he compares this relational system of understanding to Walter Benjamin's "constellation," that is, "a sort of montage in which diverse elements are brought together through the act of writing [. . .] meant to emphasize the importance of *representation* in the interpretation of history" (10). As Rothberg further explains, Benjamin coined this concept of constellation in his "Theses on the Philosophy of History," written "on the eve of 'the Final Solution'," with the aim of exposing and replacing "the concept of [linear] history and time underlying the ideology of progress," an ideology that had failed to predict or combat the forces of Nazism. As Rothberg further explains:

> Benjamin traces this failure of modernity's consciousness of time and argues that the culpable Enlightenment belief in the "irresistible" course of the "infinite perfectibility of mankind . . . cannot be sun-dered from the concept of [mankind's] progression through homoge-neous, empty time" [*Illuminations*, 260–61]. The harnessing together of different moments of time in a constellation challenges not just the "progressive" narrative form of modern history, but also its originating gesture. (10)

In the pages that follow I will attempt to demonstrate that *Dusklands* can fruitfully be analysed from the perspective proposed by Rothberg for Holocaust fiction and that the constellation of realist, modernist and post-modernist elements it displays, far from a refusal of political commitment, constitutes a clear expression of Coetzee's ideological engagement and ethical responsibility in the face of the demands of representing not only the unspeakable events of the Vietnam War and the colonisation of South Africa, but also, and most importantly, of deconstructing the originating idea of endless progress underlying the Enlightenment notion of the world as open to transformation by human intervention that has led to the jus-tification of unmitigated capitalism, colonial and imperialist domination, the discrimination and subjugation of the other, and the exploitative anni-hilation of the earth. My contention is that the novel achieves structural, thematic and ideological unity through the construction of a constellation of realist, modernist and postmodernist elements arranged in a structure of temporal regression and thematic repetition and intensification that echoes the compulsive acting out of trauma. Through repetition, the novel manages to give voice to what can be described as the ghostly trace of the founding metaphor on which the modern concept of human progress is based: the Biblical notion of *dominium terrae* (Genesis 1, 26f.), that is, the Creator's mandate to human beings to "Fill the earth and subdue it" (Genesis 1, 28), that has been described as "the cornerstone of pure 'anthropocentrism'" (Kunzmann 2005, 1 n.p.).

REALIST, MODERNIST AND POSTMODERNIST ELEMENTS IN *DUSKLANDS*

As Jane Poyner (2009) has pointed out, the publication of *Dusklands* "baffled many readers by juxtaposing two apparently discreet narratives: 'The Vietnam Project' narrated by Eugene Dawn, an American propagandist writing in the early 1970s, and 'The Narrative of Jacobus Coetzee', narrated by a line of South African Coetzees" (15). Unable to see the relation between these two narratives, most critics have tended to read them separately and also literally, without paying heed to the possibility that, as Derek Attridge forcefully argues, Coetzee's experimentalism, like that of the most innovative art of the decades between the two World Wars, might be ethical and committed, the hallmark of "an artistic practice operating under intense political pressures" (2004, 4). Attridge's view is reinforced by David Attwell's statement in the editor's introductionto *Doubling the Point* that Coetzee wrote *Dusklands* under intense political and experiential coercion, as he started conceiving the novel "in Texas during the worst years of the war in Vietnam" and he finished it in apartheid South Africa, where he decided to return after "his application for permanent residence in the United States was repeatedly denied" (1992, 9, 10). These personal circumstances create an emotional and ethical link between the apparently disparate themes of the two novellas, as Coetzee "could scarcely avoid associating the spectacle of the bombing of Vietnam with the legacy he was trying to shake off as a South African" (10).

Both novellas are presented as real-life accounts of historical episodes in modern history that had relevance both for the author and for society at the time of publication: the Vietnam War (1954–1975), which was reaching its traumatic final stage in 1974; and the contribution of one of Coetzee's ancestors to the process of colonisation that had led to the creation of the state of South Africa in 1910 and the enforcement of the apartheid regime (1948–1994), under which both the fictional translator and the flesh-and-blood author of *Dusklands* are living. This conflation of personal and general interests in the novel echoes the view, popularised by trauma critics such as Cathy Caruth or Shoshana Felman and Dori Laub, that trauma is an extraordinary external event affecting collectivities and leaving its traces on individuals. *Dusklands* enhances this dual aspect of the traumatic effects of the Vietnam War and the colonisation of South Africa by presenting the episodes of torture, subjugation and extermination of whole communities and territories from the subjective perspectives of alienated and maddened narrators suffering from perpetrator trauma. The reliability and authority

of these narrators are further called into question through the multiplication of characters called "Coetzee," the obliteration of the ontological
boundaries between flesh-and-blood and fictional personages and the fact
that the narrators often give contradictory versions of the events.

At the same time, the modernist impulse that lies behind the self-frag-
mentation and solipsistic alienation of the narrators, together with the
postmodernist multiplication of narrative instances and the obliteration of
ontological boundaries, are counteracted by the presentation of the first
novella as a scientific report on psychological warfare in Vietnam and of the
second as an academic reconstruction, edition and translation of the official
deposition of a real eighteenth-century South-African frontiersman, kept in
the archives of the Dutch East India Company at the Castle of New Hope.
These archival specifications, which correspond to the realistic impulse in
the novel, are aimed at placing The Vietnam Project and The Narrative of
Jacobus Coetzee on a par with the "objective" and "truthful" discourses of
science and history. However, this *effect de réel* is constantly undermined
by the numerous inconsistencies in the documents and the growing suspicion that the narrators are sacrificing scientific and historical objectivity
to the expression of their own strongly biased ideological outlook on the
events narrated. Thus, the credibility of the contemporary editor and of the
translator of The Narrative of Jacobus Coetzee is shattered by the sheer
bulk of their seventy-one page long translation and edition (Coetzee 2004,
51–122) of the original five-page Deposition (122n.1)—reproduced in the
novel as a three-page Appendix (123–25)—and further called into question
by the fact that this Deposition was written by "a Castle hack who heard
out Coetzee's story with the impatience of a bureaucrat and jotted down
a hasty précis for the Governor's desk" (108). The eventual discovery that
the explorer was illiterate, capable only of signing his oral Deposition with
an X (125), forcefully undermines the veracity of the coloniser's expedi-
tions, which the translator and his scholarly father had been at pains to
bring to public light. This overlapping of conflicting realism-enhancing and
realism-undermining elements produces a characteristic ambivalent effect
that begs for an ironic reading. For example, the fact that, for all his scholarly qualifications, Dr. S. J. Coetzee feels insecure about the actual name of
his eighteenth-century ancestor, whom he calls "Jacobus Janszoon Coetzee
(Coetsee, Coetsé)" (123), both undermines and reinforces his credibility
as a historian, as it uncovers a relation of hypertextuality between The
Narrative of Jacobus Coetzee and the real eighteenth-century *Journals of
Jacobus Coetsé Jansz.*[2] At the same time, the intimation that The Narrative
of Jacobus Coetzee might be a pastiche of one or more historical records
reinforces the view that Coetzee's experimentalism is aimed at breaking out
of the generic constraints imposed on transparent literature by taking the
novelistic form back to its pre-modern origins. As Gérard Genette (1997)
points out in *Palimpsests*, "hypertextuality is obviously one of the features
that enable a certain modernity, or postmodernity, to turn its back on

the age of Romantic-realist seriousness and revive a premodern tradition: *Torniamo all'antico*" (396–97). Commenting on the disturbing effect of a similar blurring of ontological boundaries and multiplication of authorial identities in Samuel Beckett's *Murphy*, Coetzee has made the cautionary remark that: "what poses as a problem for the reader of choosing rationally among authorities may be a false problem, a problem designed to yield no solution, or only arbitrary solutions" (1992, 31). If we apply this warning to *Dusklands*, we will have to conclude that the multiplication of personages and the blurring of historical and fictional records, like the juxtaposition of a mouse, Maus and Micky Mouse in Spiegelman's drawing, is not aimed at testing the readers' capacity to discriminate between authorial identities and ontologies but, on the contrary, to bring to the fore the uselessness of trying to differentiate, in this case, between fictional and historical events and personages, thus levelling history and fiction to the same category of human discourse. In this sense, the fact that the second novella is presented as an elaborate translation from the Dutch and Afrikaans into English may be said to highlight both the constructed nature of the novel and to signal language as a key aspect of the growing strain the apartheid regime was under by the middle of the 1970s. The year 1974 saw the enforcement of the Bantu Education Act that required Afrikaans, the language of the Empire, to be used on an equal basis with English as a medium of instruction. This act was met with great opposition by the young and by those sympathetic to black consciousness throughout 1975 and into 1976, leading to the Soweto students' protests on 16 June 1976 and similar outbreaks of violence throughout the rest of 1976 and 1977 (Byrnes 1996).

The hybrid fictional/historical ontology of the novel and its essentially parodic status are neatly summarised in the epigraph that precedes The Narrative of Jacobus Coetzee, which reads: "What is important is the philosophy of history" (Coetzee 2004, 53). As Attridge has noted, this epigraph directs attention to the unreliability of historical records:

> The epigraph [. . .] comes from a point in Flaubert's parodic *Bouvard et Pécuchet* when, having discovered the relatively arbitrary status of dates, the protagonists question the relevance of facts in general; this insight provides the momentary certainty of "Ce qu'il y a d'important, c'est la philosophie de l'Histoire" (190). Very soon, however, Flaubert's characters discard even this formula for other opinions. In *Dusklands*, the epigraph emphasizes the fickleness of data and directs attention to the struggle over history. (2004, 44–45)

The fact that this epigraph appears *after* the title page of the novella adds a further ironic turn to its message, suggesting that it has been written by the fictional translator or his scholarly father, not by the flesh-and-blood author of *Dusklands*. Thus, the epigraph undermines the credibility of history while at the same time its putative authors self-consciously acknowledge the fictionality of themselves and their work. The question that remains to be answered is: What is the discourse that struggles to emerge out of this self-conscious and ironic deconstruction of historical discourse? As I will try to show in the following sections, the answer to this question is to be found in the perception of the narrators' nature-hating anthropocentrism.

IRONY, ANTHROPOCENTRISM AND *DOMINIUM TERRAE* IN "THE NARRATIVE OF JACOBUS COETZEE"

Jacobus Coetzee's narrative begins with a bitter reference to the good luck of Adam Wijnand, whom he describes as the bastard son of a "Hottentot" servant who managed to establish himself in the Korana and became a rich man (Coetzee 2004, 57). This allusion is relevant not only, as Attwell cogently argues, because it helps situate the narrative within the social context of "the shift in white settlement from burgher to trekboer in the political economy of the eighteenth century" (1993, 47),[3] but also because it sets the pattern for the highly emotional style of Jacobus' narration. A similar emotionality pervades Dr Coetzee's scholarly Introduction, where, confusing history with hagiography, he presents the long forgotten coloniser as one of the earliest "heroes" who "ventured into the interior of South Africa and brought us news of what we had inherited" (Coetzee 2004, 108). This allusion to the Christian myth of origins is echoed by Jacobus' own assertion that the white colonisers differ from the natives in that: "We are Christians, a folk with a destiny." (108). Eugene Dawn, the narrator of the first novella, and the various Coetzees that participate in the writing of the second partake of this providential sense of themselves as a chosen people with a God-given inheritance. From their perspective, the ravaging of the land and the extermination or enslavement of the native peoples of South Africa respond to God's mandate to dominate the earth (Genesis 1, 26f.) and his promises to Abraham (Genesis 22, 16, 17) and David (Jeremiah 33, 17–26) that he will multiply their seed and make them the kings and priests (Revelation 1, 6) of his future kingdom on earth. This conviction allows the frontiersman to set up an essential difference between the colonisers and the Africans and to say, in words that echo the Nazis' comparison of the Jews to vermin, that the Bushman is "a wild animal, with a wild soul" (Coetzee 2004, 58), because they raid the farms during the night and disperse and mutilate the cattle as a form of revenge. Needless to say, Jacobus Coetzee is incapable of perceiving the irony of this description of the desperate Bushmen's defensive behaviour, compared to the horrifying methods used by the white farmers to drive the natives off their land, hunting them as jackals (59), with a view to exterminating or enslaving them (60); if female, the best course of action is to terrorise and rape them. Jacobus' ethical blindness in this respect is remarkable: he cannot see the dreadfulness of his methods and he in fact prefers Bushman slaves to Dutch women, because the former can be exploited sexually, without any social constraints. His conception of liberty is representative of his patriarchal ideology: he can only think of women in terms of ownership and power/bondage relationships and he has no moral scruples about the physical and psychological pain inflicted on the Bushman girls. The ethical atrophy that prevents him from perceiving the difference between affective relations, based on mutual love and equal rights, and sadomasochistic master/slave sexual practices condemns Jacobus to a terrifying solipsism that he only calls into question during his repulsive illness in the Great Namaqua camp. Confined to the menstruation hut provided by the hospitable Namaquas, the Boer explorer is prey to feverish nightmares and hallucinations during which he meditates on his life "as tamer of the wild" (77, 78), and is terrified by the possibility that there might be no life outside his psyche, that he might be dreaming himself into "a universe of which I the Dreamer was sole inhabitant" (78). However, he deliberately rejects these thoughts as an "intellectual entertainment" (78) and, after attributing them to a loss of his "sense of boundaries" provoked by the influence of the wilderness, he reasserts his status of "Destroyer of the wilderness" (79):

> The wild is one because it is boundless [. . .]. Our commerce with the wild is a tireless enterprise of turning it into orchard and farm. [. . .]. Every wild creature I kill crosses the boundary between wilderness and number. [. . .]. I am a hunter, a domesticator of the wilderness, a hero of enumeration. (80)

As this quotation makes clear, Jacobus Coetzee fully endorses the metaphysics of "boundaries" that goes back to the mythical notion of the emergence of order from chaos, whose best-known medieval expression is the *hortus conclusus*. In the book of Genesis, the primordial chaos was shaped, by the word of God, over seven days into the order we now have, and this understanding, the shaping of formlessness into form by the power of the will and the word, is crucial to the Enlightenment project of *dominium terrae*. Jacobus' obsession with counting trees and setting boundaries is representative of the Puritan mentality both of the Dutch colonisers of South Africa and of the British colonisers of New England, who are the ancestors of Eugene Dawn. These seventeenth and eighteenth-century Puritans interpreted the ordered life of village and farm, with their protective boundaries, as the manifestation of God's Kingdom on Earth, and they equated the chaotic wilderness outside with the Devil's Kingdom. From this perspective, the title of tamer or destroyer of the wilderness situates Jacobus Coetzee on a par with the Christian God of creation. Echoing the representation of a providential God as an all-seeing eye, the relentless explorer describes himself as "a spherical reflecting eye" (77) ingesting, ogre-like, the whole world into himself (79). And, using mythico-religious imagery, he describes the gun as "our mediator with the world and therefore our saviour," that is, as the (phallic) instrument used by himself and his fellow colonisers to break the primordial unity of self and world enjoyed by the Bushmen and Namaquas, and to establish an anthropocentric pattern of severance from, and domination of, the earth:

> The gun saves us from the fear that all life is within us. It does so by laying at our feet all the evidence we need of a dying and therefore a living world. I move through the wilderness with my gun [. . . and] leave behind me a mountain of skin, bones, inedible gristle, and excrement. All this is my dispersed pyramid to life. It is my life's work, my incessant proclamation of the otherness of the dead and therefore the otherness of life. (79–80)

As the allusions to the mountain of carcases and excrement and the pyramid to life suggest, Jacobus situates himself at the top a parodic *scala naturae* as sole ruler of a dead world. Mythically, the mountain, like the pyramid, is an *axis mundi* or mystical ladder connecting the three planes of existence. In poetic symbolism, the top of the mountain is the point of epiphany, "the point at which the undisplaced apocalyptic world and the cyclical world of nature come into alignment" (Frye 2001, 205). Jacobus' pyramid invokes its contrary, what Northrop Frye describes as "the dark tower, the prison of endless pain, the city of dreadful night in the desert, or, with a more erudite irony, the *tour abolie*, the goal of the quest that isn't there" (239); that is, it invokes a symbolism that denies the validity of his project to dominate the earth. In the Bible there are numerous examples of the positive and negative aspects of this symbolism. Jesus is transfigured on the top of a mount (Matthew 17, 1–9; Mark 9, 2–8; Luke 9, 28–36), and Satan falls from the pinnacle of the Temple and plunges back into darkness, while Jesus remains motionless on top of the mountain, after Satan's failure to tempt him. In Milton's version in *Paradise Regained*, Satan tempts Christ with the prospect of making him the ruler of the world, the very position Jacobus is strenuously attempting to achieve. The difference between Christ and Coetzee is that, while, "Christ discerns that Satan can never offer anything but illusions" (Russell 1990, 126), the blind frontiersman condemns himself to the infernal darkness of solipsism by failing to understand the illusory nature of his endeavour.

Jacobus' anthropocentrism stands in diametrical opposition to the holistic, nature-loving conception of life of the Bushmen and Namaquas living in the wilderness. During his first elephant-hunting expedition Coetzee is saved from sure death thanks to the hospitality of the Namaquas and the care of Jan Klawer, his devoted slave. However, once recovered from his nauseating illness (89), which fittingly mirrors his spiritual repulsiveness, he is neither thankful nor civil with slave and hosts. On the contrary, he viciously attacks a group of mocking children; he leaves Klawer to die alone on the way back to the fort;[4] and he returns to the Namaqua camp with an armed force with the only purpose of destroying the camp and all its inhabitants. This obnoxious behaviour situates him in diametrical opposition to Rebecca (Genesis 24), the Biblical figure chosen by Emmanuel Levinas to exemplify the Messianic calling to care for the other, the neighbour and the universe before looking after oneself, that takes the subject beyond both egoism and altruism to the ethical position of detachment, or emptying oneself of being.[5]

With neat accuracy, Jacobus Coetzee ends his narration justifying the massacre as part of his racial destiny and describing himself as a tool in the hands of history (106). Just as Adam's sin brought about the end of Paradise on Earth and the beginning of the Devil's Kingdom, so Jacobus' intervention in nature puts an end to the Namaquas' prelapsarian community, precipitating their fall into History. This is the sad heritage that the eighteenth-century Coetzee leaves his eager and sycophantic descendants: a barren Kingdom of spiritual darkness and physical pain ruled by master/slave relationships and the sacrifice of nature to the Enlightenment principle of endless progress that Jacobus shares with other Europeans, including the Puritan settlers that ravaged the American wilderness and exterminated the red Indians in order to create New England and leave it to Eugene Dawn as his God-destined inheritance.

IRONY, ANTHROPOCENTRISM AND *DOMINIUM TERRAE* IN "THE VIETNAM PROJECT"

As happened in "The Narrative of Jacobus Coetzee," in "The Vietnam Project" the opening words set the emotional tone of the whole novella, as it contains Eugene Dawn's account of his shocked reaction to the command of Coetzee, his supervisor, that he should revise the expert report he has just written on The Vietnam Project, a five-phase plan to destroy the Viet-cong resistance, which had already been applied by the propaganda services up to the end of Phase III. Dawn, who describes himself as a creative and brooding person who hates conflict and needs peace and love and order for his work, is so afraid of his supervisor that he admits that his first impulse was to give in and concede all in the hope that he would love him (1–2). Coetzee made his name in game theory (28), that is, in a branch of applied mathematics used in the social sciences to analyse behaviour in strategic situations, or games, in which an individual's success in making choices depends on the choices of others. This allusion to game theory refers back to the epigraph of the novella, which contains a statement by Herman Kahn (1922–1983), one of the founders of the Hudson Institute think tank and a major contributor to the analysis of nuclear warfare during the Cold War, in terms of game theory.[6] In this epigraph, Kahn defends the fighter-bomber pilots' exhilaration after succeeding in napalm-bombing their Viet-cong targets, on the reflection that it would be "unreasonable" to expect the U.S. Government to employ "pilots who are so appalled by the damage they may be doing that they cannot carry out their missions or become excessively depressed or guilt-ridden."[7] As Robert Pippin notes, "[t]he words 'unreasonable' and 'excessive' stand out in this attempt to adopt some sort of wholly analytic or objective attitude toward the moral issues involved, as if those issues could be discussed from some wholly third person and exclusively strategic point of view" (26). Eugene Dawn's split self stems from his attempt to adopt his supervisor's/Kahn's rational methods and to assume the unemotional attitude of the fighter-bomber pilots. Thus, for all his natural desire of peace and love, he entertains hair-raising ideas about how to conduct the Vietnam War and he treats his wife, a beautiful blonde, appositely named Marylin, in an abhorrent sadistic way. Just as Jacobus Coetzee preferred stunted and inert African slaves to Dutch women, so Eugene Dawn entertains necrophiliac thoughts about how he might reach ecstasy if he could make his wife sleep though the sexual act (Coetzee 2004, 12), and he understands sexual relations in power/bondage terms that presuppose the infliction of pain (10).

Marylin and her friends relate Dawn's aloofness, impotence and sadism to the horrors of war he has to delve into because of his job (10). And Marylin is convinced that her husband has a shameful secret connected with the twenty-four pictures of tortured human bodies (10) that he carries around with him all day. But Dawn rejects this view, arguing that he is not moved by the unspeakable horrors in the pictures, and he seems in fact amused by them. Thus, for example, he giggles after describing the picture of a mother carrying her son's head in a sack like "a small purchase from the supermarket" (16). This awkward response to horror is good proof of his traumatised condition since, as Freud and Breuer put forward in "On the Psychic Mechanisms of Hysterical Phenomena," psychic trauma is triggered off by the subject's incapacity to react adequately to a shocking event. Dawn also suffers from other trauma symptoms, such as sleeplessness and terrible nightmares, which add to his general state of emotional numbing, or what he himself describes as "an ecstasy of hibernation" (34). Indeed, Dawn admits that he is a sick man and, giving the Christian metaphor of the Kingdom on Earth an overtly mythical turn, he compares himself to the wounded Fisher King, whose impotence affects the fertility of the earth, reducing it to a barren wasteland (32).

However, ignoring all these symptoms, Dawn declares himself a hero of resistance to affective unsettlement, and he assumes what he considers to be his destined god-like role of imposing order over chaos by strictly rational means (27). Echoing the Biblical mandate to dominate the earth, he uses prophetic language to describe his messianic calling to save his countrymen, and, just as Jacobus Coetzee justified the massive killing of animals and human beings as proof of the existence of a world outside his own psyche, he sees the massive killing of Viet-cong enemies as the result of the U.S. tragic need to transcend their nightmarish solipsism: "Our nightmare was that [. . .] we did not exist; that, since whatever we embraced wilted, we were all that existed" (17). Dawn's entropic fear of inexistence, like that of the Boer frontiersman, is symptomatic of his unmitigated anthropocentrism. In the case of Dawn, this aspect of his psychic imbalance is given a self-conscious, postmodernist turn by his constant references to the act of writing and his shattering suspicion that he is a puppet in the hands of Coetzee, his supervisor, whom he describes as an all-controlling and awe-inspiring, though creatively diminished, authorial "manager" (1). This perception of himself problematises his ontological status, already called into question by the curt remark that opens his narration: "My name is Eugene Dawn. I cannot help that. Here goes" (1). This remark points to the narrator's awareness that his name is preternaturally appropriate for someone working on the Vietnam project, as it offers an (ironic) counterpart to the darkness of the times evoked by *Duskland*s. As Dominic Head argues, the title evokes "the encroaching historical night of the interior 'dusklands'" in opposition to "the idea of a new dawn of historical hope—as a literary trope" (168n19). Needless to say, this trope refers to modernity's notion of history as endless progress, the metaphor on whose deconstruction the whole novel develops.

From an intertextual perspective, it is easy to see that the name and behaviour of Eugene Dawn parody those of Eugene Gant, the protagonist of Thomas Wolfe's 1929 autobiographical novel, *Look Homeward, Angel: A Story of the Buried Life*. Like Dawn, Gant is a U.S. American who lives through a dreadful armed conflict, in his case, the First World War, without actually participating in it and, also like Dawn, he is so exultant about the possibilities of a new historical dawn opening up after it, that he earns a reputation of being mad. Also like Dawn, he thinks of himself as the ruler of the world. But where Dawn's psychical trauma is the result of his anthropocentrism, Gant's "madness" is the product of his romantic attachment to nature, his relish for sexual love, and his unbounded faith in seasonal (and therefore historical) renewal.

Unlike Gant, who basks in the sun and gathers energy from his close contact with nature and the fruition of love, Dawn prefers the "fatherly" hum of the library lights and he finds himself in paradise surrounded by books (Coetzee 2004, 7). However, for all the bliss of this artificial heaven, he is unable to carry on creative work in the library and he admits that he can only write about Vietnam in the early hours of the morning, facing the rising sun, "when the enemy in [his] body is too sleepy to throw up walls against the forays of [his] brain" (6). The ironically named, sad expert in psychological warfare feels an ardent desire to help his fellow countrymen move beyond the darkness of the times to a new "dawn" of historical hope. Thus, facing the rising sun for inspiration, he strenuously works to apply the rational principles of game theory to the situation in Vietnam, but he is constantly betrayed by his revolting body, which interrupts his intellectual work with headaches, yawns, spasms and other bodily symptoms, climaxing in the emergence of an "inner face" with "monstrous troglodyte features" (7). The emergence of this troglodyte face associates Dawn with the protagonist of Nikolai Gogol's satiric story, "The Nose" (1836), and with many other self-fragmented characters suffering from neurosis, psychic dissociation and paranoia. The revolt of his body neatly replicates the mythical scheme Dawn employs to explain the Vietnam War to the military, which is the Greek myth of the ever-recurrent conflict between Father and Sons, with Mother Earth on the side of the rebellious children. Thus, he presents the Viet-cong as the rebellious sons and the United States as the tyrant-father, and he describes the possibility of the sons' victory as "a humiliating blow that renders him [the father] sterile" (26). The solution Dawn proposes to put a definitive end to this ever-recurrent pattern of rebellion is the annihilation not only of the malcontent children but of their scheming mother as well (29). Dawn's preposterous scheme for total annihilation of the earth, like Swift's "Modest Proposal" to get rid of pauper children by turning them into mince pies to feed the rich, has the incontestable force of pure rationality deprived of the burden of an ethic of affect. With this scheme, the Enlightenment idea of historical progress and the Biblical mandate to dominate the earth reach their logical, apocalyptic end. The problem is that the price Dawn has to pay for devising this scheme is madness.

According to Freud and Breuer, psychic trauma can be produced by "[a]ny experience which calls up distressing affects—such as those of fright, anxiety, shame or physical pain, including the moral injury inflicted by the ill-treatment of a superior or the witnessing of the pain of a beloved one" (14). As they explained, the trauma is not caused by the crudity of the event itself but wholly depends on *"whether there has been an energetic reaction to the event that provokes an affect. [. . .] If there is no such reaction, whether in deeds or words, or in the mildest cases in tears, any recollection of the event retains its affective tone to begin with"* (8). This description of psychical trauma perfectly fits Dawn's fear of confrontation with his supervisor, his growing sexual apathy, aloofness and tendency to hibernate. As readers learn at the end of the novella, Dawn tries to overcome this incapacitating paralysis by undertaking a rash action: he kidnaps Martin, his eight-year old son, and flees with him to his native state of California with the unrealistic aim of starting a new life at the fittingly named "Loco Motel" under the assumed names of "George Dobb and son" (Coetzee 2004, 35). The choice of this pseudonym reinforces the link between Dawn's theories and the Enlightenment, pointing as it does to Dawn's admiration for Maurice Dobb (1900–1976), the British economist known for his Marxist interpretation of neoclassical economic theory.

After kidnapping his son, Dawn feels exhilarated since he interprets his crazy action as proof that he has overcome his physical paralysis and emotional numbing (35). But the truth is that, at this stage, he is irrecoverably immersed in his own psychotic world. The novella ends with Dawn in a mental institution undergoing a psychoanalytical treatment that includes the interruption of sleep because it "stimulates dreaming and facilitates recall" (49). What he remembers by these means sets his perpetrator trauma within the larger context of an earlier, latent family trauma transmitted to him by his vampiric mother and soldier father, itself the symptom of the collective trauma brought about by the colonisation of America:

> There is still my entire childhood to work through before I can expect to get to the bottom of my story. My mother (whom I have not hitherto mentioned) is spreading her vampire wings for the night. My father is away being a soldier. In my cell in the heart of America, with my private toilet in the corner, I ponder and ponder. I have high hopes of finding whose fault I am. (49)

CONCLUSION

As I hope to have shown, finding whose fault he is is the question J. M. Coetzee set himself to answer by writing *Dusklands*. Rejecting the progressiveness of Enlightenment history and Biblical destiny, Coetzee arranges the novel in a structure of repetition and variation comparable to a Benjaminian constellation, presenting the ideal of *dominium terrae* as the collective acting out of a perpetrator trauma provoked by atrocious acts of colonisation, subjugation and destruction of the radical other, initiated in the eighteenth century and re-enacted once and again throughout the modern period. This structure of repetition-*cum*-variation begs for a circular as well as a backward reading, from the 1760 Deposition of Jacobus Coetzee to the 1974 report on The Vietnam Project. Thus, the three-page eighteenth-century account of the crossing of a natural frontier in a foreign land is repeated and enlarged by the various twentieth-century narrators into the evermore hallucinatory rendering of the ultimate consequences of accomplishing the Biblical mandate to dominate the earth. By tracing the origin of the collective trauma of colonisation back to the Enlightenment values of rationalism, individualism, Utilitarianism and endless progress, J. M. Coetzee effectively deconstructs the progressiveness of World History through empty time, bringing to the fore the evils of anthropocentrism in contrast to the nature-loving and communal values associated with the Viet-cong peasants in the first novella and with the Bushmen and Namaquas in the second.

*The research carried out for the writing of this article is part of a project financed by the Spanish Ministry of Economy and Competitiveness (MINECO) (code FFI2012–32719). The author is also grateful for the support of the Government of Aragón and the European Social Fund (ESF) (code H05).

NOTES

1. Granofsky places Coetzee at the centre of what he describes as "the vogue after World War Two of rewriting classic texts [in order to] make clearly discernible subtexts that were previously invisible [. . .] in that they reflected the dominant ideology of their culture" (156).

2. According to Dominic Head, Coetzee employed the following eighteenth-century sources: "*The Journals of Brink and Rhenius*, translated by E. E. Mossop (Cape Town: The Van Riebeeck Society, 1947), and *The Journal of Hendrik Jacob Wikar, and the Journals of Jacobus Coetsé Jansz, and Willem van Reenan*, translators A. W. van der Horst and E. E. Mossop (Cape Town: The Van Riebeeck Society, 1935)" (1997, 167n4).

3. Attwell's argument was made in response to the complaints of inaccuracy in Coetzee's rendering of the story of Adam Kok, which, according to Peter Knox-Shaw, provided the background to Adam Wijnand's story. Knox-Shaw is representative of the large number of realism-biased critics who found fault with Coetzee's use of sources in his South-African novels in the 1970s and 1980s (47). See Knox-Shaw (1983, 1, 22–24, 65–81). For a discussion of this issue, see Head (1997, 29, 167n4).

4. Jacobus gives two different versions of his slave's death; in the first Klawer was swallowed by a hippopotamus hole (94). In the second, he caught a cold and was left by Jacobus in a little cave, with the promise that he would return with help (95). The incompatibility of these versions enhances the unreliability not only of the narrator but also of the contemporary editor. As Attridge observes, this and other "glaring inconsistencies among the three accounts of Jacobus's expedition" are examples of "the undermining of documental verisimilitude" (16).

5. See Chalier (1991) on this.

6. Two of Khan's most controversial books are *On Thermonuclear War* (1960), a treatise on the nature and theory of war in the thermonuclear age, where he used game theory to discuss the strategic doctrines of nuclear war and its effect on the international balance of power; and *Thinking About the Unthinkable in the 1980s* (1984), where Kahn calmly analysed in brutal detail the horrendous forms a post-nuclear world might assume.

7. As David Attwell has pointed out, this epigraph belongs in a set of real documents on the Vietnam War gathered together and printed by the Hudson Institute in 1968 in a book entitled *Can We Win the Vietnam War: A Dilemma* (1992, 40).

WORKS CITED

Adorno, Theodor W. (1955) 1981. *Prisms*.Translated by Samuel Weber and Shierry Weber. Cambridge, Massachusettes: MIT Press.

Attridge, Dereck. 2004. *J. M. Coetzee and the Ethics of Reading*. Chicago and London: The University of Chicago Press.

Attwell, David. 1992. "Editor's Introduction." In Coetzee, 1992, 1–13.

———. 1993. *South Africa and the Politics of Writing*. Berkeley, Los Angeles, Oxford: University of California.

Byrnes, Rita M., ed. 1996. "Soweto, 1976." In *South Africa: A Country Study*. Washington: GPO for the Library of Congress. Accessed January 9 2011. http://countrystudies.us/ south-africa.

Caruth, Cathy, ed. 1995. *Trauma: Explorations in Memory*. Baltimore and London: Johns Hopkins University Press.

Chalier, Catherine. 1991. "Ethics and the Feminine." In *Re-Reading Levinas*. Edited by Robert Bernasconi and Simon Critchley, 119–29. London: Athlone.

Coetzee, J. M. 1970. "The Comedy of Point of View in Beckett's *Murphy*." In Coetzee 1992, 31–38.

———. 1992. *Doubling the Point: Essays and Interviews*. Edited by David Attwell. Cambridge. and London: Harvard University Press.

———. (1974) 2004. *Dusklands*. London: Vintage.

Felman, Shoshana and Dori Laub. 1992. *Testimony: Crises of Witnessing in Literature, Psychoanalysis, and History*. New York and London: Routledge.

Freud, Sigmund and Josef Breuer. (1893) 2001. "On the Psychical Mechanism of Hysterical Phenomena: Preliminary Communication." In *Studies on Hysteria, The Standard Edition of the Complete Psychological Works of Sigmund Freud*, Vol. II. 1893–1895. Edited and translated by James Strachey, Anna Freud and Alan Tyson, 1–17. London: Vintage.

Frye, Northrop. 1957. *Anatomy of Criticism: Four Essays*. Princeton, New Jersey: Princeton University Press.

Genette, Gérard. (1982) 1997. *Palimpsests: Literature in the Second Degree*. Translated by Channa Newman and Claude Doubinsky; foreword by Gerald Prince. Lincoln: University of Nebraska Press.

Granofsky, Ronald. 1995. *The Trauma Novel: Contemporary Symbolic Depiction of Collective Disaster*. Bern: Peter Lang.

Haeming, Anna. 2009. "Authenticity: Diaries, Chronicles, Record and Index-Simulation." In *J. M. Coetzee in Context and Theory*. Edited by E. Bochmer, R. Eaglestone, and K. Iddiols, 173–84. New York and London: Continuum.

Head, Dominic. 1997. *J. M. Coetzee. Cambridge Studies in African and Caribbean Literature*. Cambridge: Cambridge University Press.

Knox-Shaw, Peter. 1983. "*Dusklands*: A Metaphysics of Violence." *Commonwealth Novel in English* I (2): 65–81.

Kunzmann, Peter. 2005. "Biblical Anthropocentrism and Human Responsibility." Konferencja Chrzescijanskiego Forum Pracowników Nauki *Nauka—Etyka—Wiara 2005*, 1–5. Accessed February 16 2011. http://chfpn.pl/files/?id_plik=272.

Levinas, Emmanuel. 1969. "Judaism and the Feminine Element." *Judaism*, XVIII (1): 33–73.

———. (1974) 1981. *Otherwise than Being: or, Beyond Essence*. Translated by Alphonso Lingis. The Hague: Martinus Nijhoff.

Pippin, Robert. 2010. "The Paradoxes of Power in the Early Novels of J.M. Coetzee." In *J. M. Coetzee and Ethics. Philosophical Perspectives on Literature*. Edited by Anton Leist and Peter Singer, 19–41. New York: Columbia University Press.

Poyner, Jane. 2009. *J. M. Coetzee and the Paradox of Postcolonial Authorship*. Farnham, Surrey and Burlington, Vermont: Ashgate.

Rothberg, Michel. 2000. *Traumatic Realism: The Demands of Holocaust Representation*. Minneapolis and London: University of Minnesota Press.

Russell, Jeffrey Burton. (1986) 1990. *Mephistopheles: The Devil in the Modern World*. Ithaca, New York: Cornell University Press.

Wolfe, Thomas. (1929) 1957. *Look Homeward, Angel: A Story of the Buried Life*. New York: Scribner.

14 "There's that curtain come down"

The Burden of Shame in Sarah Waters' *The Night Watch*[*]

Maite Escudero-Alías

After three much-praised and successful novels about Victorian under-worlds and young women discovering and redefining their sexual identities, Sarah Waters' *The Night Watch* (2006) is set in the bombed-out London of the 1940s during the Second World War. In spite of the abrupt change of historical setting, there are certain features in this novel, such as its urban context, the interest in class, the depiction of unmarried and child-less women as "social crimes" (Llewellyn 2004, 207) or the use of spectral imagery, which are recurrent in her fiction. Yet, the focus on the devastating atmosphere of the war and its aftermath, combined with a thorough exami-nation of individual traumas as veiled emotions that lead to a devastating psychic life in their protagonists, evoke an engagement with longstanding critical thought regarding both trauma studies and its discontents.

Moreover, this sense of emotional failure is prompted by the novel's struc-ture: apparently more austere in style and tone—it is narrated externally—*The Night Watch*'s structural vehicle is a reverse chronology that recedes from the exhausted present of 1947 back through the intense bombings of 1944 to the apocalyptic atmosphere of 1941. Often used in trauma fiction as a way of placing the reader in a timeless, chaotic pattern of existence, this reverse chronology also enacts here an ethical form of telling which is relevant to queer and minority cultures. Significantly, in *The Night Watch* the narrative moves backwards, as if in an attempt to "produce alternative temporalities," to use Judith Halberstam's words, that "allow their par-ticipants to believe that their futures can be imagined according to logics that lie outside of those paradigmatic markers of life experience—namely, birth, marriage, reproduction and death" (2005, 2). Thus, the reader dis-covers through the eyes of the queer characters an increasing atmosphere of emotional upheavals and unbearable secrets, which bring about a genu-ine dramatic plot enhanced by their experience of wartime London, "a city being blown and shot to bits" (Waters 2006, 411). The queering of space and time encompasses a potentiality not only to rewrite literary and cultural representations of different communities and individuals but also to imagine their lives differently. Besides this narratological device of new temporal logics, Waters makes use of an overt manoeuvre which is relevant to my purposes here: by giving agency to homosexual characters her fiction also contributes to moving beyond heteronormative master narratives of Western culture.

Waters' search for alternative models of literary representation especially within the first decade of the twenty-first century, may spur us as literary critics to accept the challenge and responsibility of delving into new critical and discursive paradigms along current premises of thought. It is precisely in this spirit of moving beyond conventional epistemological venues of representing and understanding trauma that the present contribution attempts to expand its theoretical and analytical scope. For this purpose, I shall concentrate on an exploration of the affect of shame as it appears in the novel, and this for several reasons. Firstly, because shame has been all too often disregarded as a worthless affect by both psychoanalysis and trauma studies; secondly, because, unlike other affects like guilt or disgust, shame attacks the global system of self-esteem and points to an individual and social failure of the person who suffers it, and lastly, I will refer to a type of invisible shame which, due to its structural resemblance with melancholia, recalls a highly painful affective reaction because it "involves painful self-scrutiny, and feelings of worthlessness and powerlessness" (Lewis 1992, 123). Although these are universal human feelings that most people suffer occasionally, the effects of heterosexism, homophobia and racism often contribute to a more negative topography of affects and emotions. As Sara Ahmed has argued, "such forms of discrimination can have negative effects, involving pain, anxiety, fear, depression and shame, all of which can restrict bodily and social mobility" (2004, 154). Not coincidentally, there is in *The Night Watch* a melancholic shame that lies beneath and is permanently attached to the protagonists' queer identities since, as mentioned above, the plot is woven around a burden of shameful secrets that turn out to be unbearable for each and all of the protagonists.[1]

In examining the affect of shame as a narrative of (im)possibility within the field of trauma studies, the present chapter also draws upon interdisciplinary theoretical insights that seek to challenge classical psychoanalytic accounts of affects, mourning and melancholia: namely, from Silvan Tomkins' revolutionary theory of affects, coined in the 1950s, through Nicholas Abraham and Maria Torok's rewriting of Freudian psychoanalysis with their concepts of the phantom, introjection and incorporation, up to more recent elaborations of affects not only as powerful spatial emotions "effecting displacement and effacement in its subjects" (Munt 2008, 80) but also as structures of feeling which may incorporate a productive account of personal, communal and national identities (Sedgwick 2003; Ahmed 2004, 2010).

Indeed, the consideration of shame as the affect of "indignity, of defeat, of transgression, and of alienation" (Tomkins 1995a, 133) and its potentiality to be transformed into positive affects were thoroughly theorised by Silvan Tomkins in his groundbreaking work *Affect Imagery Consciousness* (1962–1992).[2] Before Tomkins' work, shame was basically considered a visible emotion that could easily be identified by its physical manifestations. Thus, from Charles Darwin to Sigmund Freud, the study of shame had been reduced to highlighting its physiological traits. In his text *The Expression of Emotions in Man and Animals* (1872) Charles Darwin offered a pioneer physiological description of shame as a blushing of the face or an averted gaze with a "strong desire for concealment" (Darwin 1999, 319). Sigmund Freud, on the other hand, overlooked the importance of affects in the formation of the subject and considered shame as a reaction formation whose function was to maintain the repression of forbidden exhibitionistic impulses, as a kind of moral watchman "who maintains repressions [. . .] as a dam against immoral, exhibitionistic excitement, organically determined and fixed by heredity" (1905, 138). Freud's premise that such moral dams are genetically determined—restricting human dignity and emotions to basic drives of survival—entails serious limitations to his own theories and in particular to his theory of free will, a freedom that rises from the complexity of the human affect system itself. Moreover, Freud's overestimation of guilt vis-à-vis shame has historically relegated the latter to academic and social oblivion. So, whereas guilt attaches to what one does, therefore signalling a specific experience, shame attaches to and sharpens the sense of what one is, enacting an ontological and phenomenological topography which can last a lifetime.

As a response to Freud's trivialisation of the affect system, Tomkins elaborated on this theory of affects and, particularly, on the affect of shame. In contrast with the Freudian centrality of drives, Tomkins' most valued contribution lies in the primacy of affects that he sees as the main motivational system of human life. Thus, unlike drives, which are more limited in time and in density due to their instrumentality, affects can be endowed with a greater motivational freedom involving an extraordinary competence to develop cognitive and emotional learning. Another key distinguishing feature is that whereas drives are primarily self-fulfilling, affects can be autotelic; that is, self-reinforcing agents with essentially affective responses. This means that the affect system has both self-rewarding and self-punishing characteristics that can be temporarily transformed into negative and positive responses respectively. Accordingly, any negative affect can be invested in another positive affect and is also capable of restructuring the emotional charge of such effect. In Tomkins' words:

> Affects may also be invested in other affects, combine with other affects, intensify or modulate them, and suppress or reduce them. In marked contrast to the separateness of each drive, the emotions readily enter into combinations with each other and readily control one another. (1995b, 56)

The potential of affects to be attached and connected to other affects provides a new epistemological framework from which to counteract the negative and pernicious effects of personal and collective traumas. Not coincidentally, Tomkins' theories have been thoroughly analyzed by literary and cultural critics such as Eva K. Sedgwick (1995, 2003), Sara Ahmed (2004, 2010) and   m have focused on the malleability of sh   a promising structural and semantic free   . Sedgwick, for instance, insists on how the predominant role of shame over other affects in the formation of identity and its potential to alter pathological accounts of minority identities based on their race, sexuality, class or nationality must be seen as meaningful clues possessing a contingent orientation towards the healing and well-being of traumatised individuals and communities. Likewise, Sally Munt theorises on the notion of shame as produced by the circulation of different emotions that can be easily attached to "envy, hate, contempt, apathy, painful self-absorption, humiliation, rage, mortification and disgust" (2008, 2). Furthermore, she considers shame a mobilising agent of the self and communities that "can also produce a reactive, new self that has a liberatory energy" (80). As will be argued, if the forms taken by shame are available for the work of metamorphosis and transfiguration, a distinction can be drawn, then, between shame as a monolithic entity that prevents any productive effect on trauma and as a malleable affect that can be turned back against itself to re-create a new narrative of recovery.

Following Tomkins' consideration of shame as a permanent, structuring axis of identity formation—i.e. due to its early appearance in the infant's psychosexual development, shame is crucial in all identity formation—I would like to emphasise the centrality of shame in the traumas related to the configuration of minority identities. As a keystone affect, then, shame is largely connected with the whole feeling about oneself and it brings into focus questions such as those of self and identity or feelings of inferiority and failure. If, as most critics agree, shame constitutes "a failure of the ego to reach a narcissistic ideal" (Pajaczkowska and Ward 2008, 5), my contention here is that the pain of shame in queer persons can be intensified by their failure to reach the heterosexual ideal and follow heteronormative lifestyles. Admittedly, being queer is also to be positioned on the margins of social and cultural recognition with the subsequent psychic cost that this implies; that is, persons who live a life that does not comply with heteronormative practices are more prone to suffer states of melancholia, pain, depression and shame than others who follow the linear axes of birth, marriage, reproduction and death. Sara Ahmed has repeatedly argued that the feeling of being outside the limits of cultural and social intelligibility often brings about both a personal alienation and a feeling of disorientation that encapsulate a "politics of discomfort" because "heteronormativity functions as a form of public comfort by allowing bodies to extend into spaces that have already taken their shape" (2004, 148). Such an awkward positioning, that is, the position of being queer, reminds us that life is not always linear and that there are certain spatial restrictions which can mark a failure to reproduce the expected desires, affects and emotions. The affects resulting from not following the right paths, from "loving a body that is supposed to be unlovable for the subject I am" (Ahmed 2004, 146) are multiple, among which shame and melancholia are always included (Braidotti 1993; Butler 1997).

In her elaboration of a queer phenomenology, Ahmed (2006) emphasises that following lines of resistance and rebellion against heteronormativity may give rise to new impressions of reality. Thus, for a lesbian to live a disoriented life does not exclusively mean inhabiting a body that is not recognised by the social and political order, but it also means embodying a politics of hope which might generate new paths of desire and different routes. Even though maintaining a permanent positive transgression is not psychically, socially and materially possible for some individuals, Ahmed points out that the ambivalence of these discriminatory structures may indeed allude to a self-rewarding component because "the non-fitting or discomfort opens up possibilities, an opening up which can be difficult but exciting" (2004, 154). Just as the feminist movement has been historically peppered by negative affects such as anger, humiliation or pain transforming them into love, hope and courage, so, Ahmed states, that racial and sexual minority cultures should reinforce creativity and dynamism to articulate more hopeful affective attachments. Given the malleability of affects, then, lesbian bodies can be reoriented towards unofficial yet fulfilling paths of love, hope and endurance that may counteract the pernicious effects of shame.

Accordingly, Waters' *The Night Watch* is imbued with a set of elements which attest to the double-edged behaviour of shame: as a stigma, it represents a failure or absence of contact, indicating personal and social isolation; and yet, the devastating effects of shame can be neutralised by resetting affects such as joy, excitement, interest or love. Consistently associated with motifs of invisibility and secrets, the four protagonists of this novel share a shameful queerness which stems from their own identities as sexual and social outsiders. We first meet Kay, a mannish lesbian "regularly called 'young man', and even 'son'" (Waters 2006, 5) because of her trousers and masculine hair-cut: "don't you know the war's over?" (94), people ask her, for the war had been her finest hour, as an ambulance driver, making her feel "awake, alert, alive in all her limbs" (179), even at the expense of collecting "unidentified bits of body parts, feeling the ghastliness of them, the awful softness of human flesh, the vulnerability of bone, the appalling slightness of necks and wrists and finger-joints" (200). Once the war is over, Kay has nothing to do but "walk restlessly for hours at a time—stiller than a shadow, because she'd watch the shadows creeping across the rug" (4), contemplating the loss of love and purpose in her life. The depiction of Kay as a butch lesbian who "haunted the attic floor like a ghost or a lunatic [. . .] dissolving into the gloom which gathered, like dust, in its crazy angles" (4) corroborates the spectral imagery with which lesbians have been traditionally stereotyped. Significantly, the association of lesbianism with spectral and non-human figures has been widely documented by Terry Castle in her outstanding work *The Apparitional Lesbian: Female Homosexuality and Modern Culture*. As she puts it:

> The lesbian remains a kind of "ghost effect," elusive, vaporous, difficult to spot [. . .]. The lesbian is never with us, it seems, but always somewhere else: in the shadows, in the margins, hidden from history, out of sight, out of mind, a wanderer in the dusk, a lost soul, a tragic mistake, a pale denizen of the night. (1993, 2)

The reduction of lesbianism to non-existence and invisibility does propel the affect of shame into an epistemological void, which resembles a self-consuming and faded condition of the self. By drawing a parallel between lesbianism and shame as both invisible and unreadable, the lesbian becomes the emblem of a threatening phantom caused by the shameful condition of being a lesbian. For Abraham and Torok, the concept of the phantom is imbued with a collective imagery that is beyond the individual. This means that "the symptoms do not spring from the individual's own life experiences but from someone else's psychic conflicts, traumas or secrets" (1994, 166); that is, the phantom is generationally transmitted and comprises not only familial secrets, but also communal and even national, political and social shames. If, as Abraham and Torok have argued, the phantom also represents "the interpersonal and transgenerational consequences of silence" (168), we should look for ways of reducing it, or else, of diminishing its pernicious effects. Following this line of argument, I would venture to say that the existence of a lesbian phantom, understood as a collective entity that haunts Kay and drags her into devastation and solitude, may subtly "counteract libidinal introjection" (174), which is to say that this phantom coincides with Freud's description of the death instinct:

> First of all, it has no energy of its own; it cannot be "abreacted", merely designated. Second, it pursues its work of disarray in silence. Let us note that the phantom is sustained by secreted words, invisible gnomes whose aim is to wreak havoc, from within the unconscious, in the coherence of logical progression. Finally, it gives rise to endless repetition and, more often than not, eludes rationalization. (Abraham and Torok 1994, 175)

This description fits Kay very well: she becomes the recipient of social injustice, shameful secrets and nameless suffering, in spite of her status as an upper-class lesbian. Indeed, to claim that a lesbian phantom is the genesis of Kay's shame also suggests that only by unveiling the lesbian phantom as such can silence be broken and shame deciphered. Furthermore, if the phantom is disclosed and revealed as a conscious element that is not alien to the subject who harbours it, then, specific prohibitions, phobias and fears can be released. Yet, the tasks of disclosing one's own secrets and counter-acting the phantom seem impossible in this narrative, especially for lonely women who must learn to live with death, both literally and metaphori-cally. As Kay says to her butch friend Mickey, with whom she used to work in the ambulance, "Look at me, Mickey! Look at the creature I've become! Did we really do those things we did?—you and I, when the war was on? [. . .] I can't get over it, Mickey, I can't get over it" (Waters 2006, 101).

Kay's drained existence in the aftermath of the war trauma is drastically marked by the betrayal of her former lover, Helen, with Kay's close friend Julia, during the war. The loss of Helen is ironically portrayed through an exquisite *tour-de-force* in which Kay, believing that their place has been bombarded as she listens to the terrible explosions on the ambulance's radio, and appalled by a mixture of pain and fear, drives there and finds Helen and Julia, standing a little way off in the street. When embracing Helen, Kay

> didn't feel pleasure or relief. She felt only, still, a mixture of pain and fear so sharp, she thought it would kill her. She shuddered and shud-dered, for she could understand nothing, at that moment, except that Helen had been taken, and now was returned. "Julia. Oh Julia. Thank God! I thought I'd lost her". (Waters 2006, 424)

Indeed, Kay loses Helen, not due to the intense bombings of that night, but because she is having an affair with Julia. By having lost "the only thing that makes this bloody war bearable" (305), Kay's days become blank and hopeless. The personal and social cost of loving a same-sex person is often evoked in this novel as an invisible source of shame and despair: not only can Kay not "take Helen in her arms" when walking together in the park, as men would do, but the "consciousness that she must not do it" (305) makes her ill. This feeling of unbearable grief because of being "fucking eunuchs upstairs" (46) is shared by Helen, Kay's ex-girlfriend and now in love with Julia, an upper-class fiction writer.[4]

Actually, the psychic cost of loving a same-sex person is shown in Helen who is portrayed as being mad, obsessive and paranoid, for she attempts to commit suicide when her relationship with Julia starts foundering. Helen's passionate love towards Julia—"*I'm in love with Julia! It's a marvellous thing, but terrible too. Sometimes it feels like it's almost killing me! It leaves me helpless. It makes me afraid! I can't control it!*" (Waters 2006, 109)—is gradually turned into madness and despair, for Helen's unjustified jealousy ends up killing their relationship. Yet, such a secret love that "is almost killing Helen" (113) constitutes both the only source of pleasure and joy—i.e. the resetting of positive affects through which to undermine the mortifying features of shame—and the crypt, to use Abraham and Torok's concept, in which silence and untold secrets become major obstacles to the individual's emotional development and self-expansion. According to Abraham and Torok, silence and its varied forms work to block introjection, "the process of psychic nourishment, growth, and assimilation [. . .] that allows human beings to continue to live harmoniously in spite of instability, devastation, war, and upheaval" (1994, 14). Drawing upon Freud's principles of abreaction and working-through, these authors contend that suffering and traumas that are recognised as such can be assimilated and modified through psychic introjection. Unlike the concepts of "crypt" and "incorporation", which denote the individual's refusal to introject loss, thus creating a psychic tomb in which the shameful secret is kept and covered up, introjection is "a broadening of the ego through transferential love" (Abraham and Torok 1994, 127). The guiding principle for the continual process of self-recovery is introjection; that is, the idea that the self is in a constant process of growth, open to change and metamorphosis. Or, to put it in simpler terms, only if the phantom is reduced, can secrets and negative affects be defeated. In Abraham and Torok's terms: "Reducing the 'phantom' entails reducing the sin attached to someone else's secret and stating it in acceptable terms so as to defy, circumvent, or domesticate the phantom's (and our) resistances, its (and our) refusals, gaining acceptance for a higher degree of 'truth'" (189). However, neither Kay nor Helen are able to overcome the suffering from their encrypted shame for, as has been mentioned earlier, inexpressible secrets such as one's lesbianism can lead to the erection of an internal tomb which can gradually kill the person. As the heterodiegetic narrator claims:

> She [Helen] could never convey to Julia how utterly dreadful it was to have that seething, wizening little gnome-like thing spring up and consume you; how exhausting, to have to tuck it back into your breast when it was done; how frightening, to feel it there, living inside you, waiting its chance to spring again. (Waters 2006, 147)

By turning her love towards Julia into an obsessive and self-destructive source of jealousy and hate—"*What an idiot you are! Julia loves you. It's only this beast in you she hates, this ridiculous monster*" (57)—Helen misses the opportunity of overcoming her shameful thoughts. As a matter of fact, she attempts to commit suicide and fails, only to increase her embarrassment and shame.

Helen's inability to transform her shame into blissful affects is also embodied by Vivien, Helen's workmate at a lonely-hearts bureau. In spite of providing hope and healing for heart-broken people in the midst of a destructive war, both characters are endowed with a stark crypt, which damages "their souls, their more inner sense of identity and humanity" (Budden 2009, 1034). Viv is the only character that is not queer sexually: she has an affair with a married man called Reggie, gets pregnant and then is compelled by her lover to have an illegal abortion with serious complications. As a result, she almost dies and it's only Kay's quick assistance with the ambulance and her sense of complicity when she realises that Viv is a single woman that saves her life. Viv perceives Kay's assistance as a sublime attempt to bring beauty into war and chaos, for Kay was "terribly kind. Wartime is a time of kindness. We all tend to forget [. . .] the courage of people, the impossibility of goodness" (Waters 2006, 130). In spite of this altruistic act of goodness in the midst of destruction and void, the truth is that Viv's life resembles and merges into the hardness, dryness and filth of London during the war. Not coincidentally, Viv's pregnancy and relationship with a married man constitutes her shameful secret, so much so that "the show of bitterness, the blood of self-pity, had worn her out [. . .]. The pain and the pain were utterly black, and timeless" (372, 381). This shameful fact will be encrypted in Viv's heart forever, signaling "something disappointed about her, a sort of greyness. A layer of grief, as fine as ash, just beneath the surface" (17). So, the expression "there's that curtain come down" (17), uttered by several protagonists throughout the novel, becomes a recurrent metaphor here for Viv's covered shame, an invisible shame which would, nevertheless, "kill her father, *after everything with Duncan*" (276), that is, Viv's brother.

Duncan, "who had some queerness or scandal attached to him" and "who saw himself as a kind oddity or fraud" (Waters 2006, 17, 89) is portrayed as a queer man fascinated by antiques, old beyond his years, living in a suburban seclusion with a mysterious uncle, wincing from his past, a bunch of hidden desires and panic attacks. Thus, as the plot of the novel unravels retrospectively, the reader comes to know about Duncan's source of panic and shame, for he spent some years in prison, accused of having murdered his lover Alec just before the war. Indeed, Duncan and his friend Alec were planning to commit suicide as a consequence of their conscientious objection to the war. In the end, Duncan could not do it and, as their secret became unveiled, he had to bear the enormous guilt and shame of his years in prison, conscientious objection and suicide plan. Duncan's pervasive feelings are those of shame and disappointment in his father, in himself and even in his sister Viv. He blushes too easily in front of the others, and his shame "has drained him, emptied him out" (91), as if it were a fragmentation of consciousness, an emotional dislocation prompting self-blaming devaluations and persistent feelings of inadequacy and failure: "his disappointment would start to rise up and almost choke him [. . .] Am I ashamed! Of course I'm bloody ashamed! You don't know what shame feels like, in here" (309, 311).

These utterances imply both recognition of his shame and a plea for empathy and understanding. Of course, such grief reflects Duncan's negativity and discomfort towards others, occupying a queer phenomenological position that defines him as abject, as "cast out from the domain of the liveable" (Butler 1993, 9) or even as the mysterious man who, in failing to reproduce heteronormativity, becomes a member of the living dead, "making himself be stuck, as a way of—punishing himself, for all that happened years ago, all that he did and didn't do" (Waters 2006, 119). Equally though, Duncan's inability to overcome grief is commented on by Viv, as if it were a melancholic state of self-denial and emotional blockage that prevents them from being outstandingly good: *"This is what happens to people like me. I'm just like Duncan, after all. We try to make something of ourselves and life won't let us, we get tripped up"* (270). In fact, the definition of Viv and Duncan—and by extension of Helen and Kay—as ghostly dead characters also recalls Abraham and Torok's premises according to which prohibitions or the secret of living a queer life cause individuals to be caught between two inclinations difficult to reconcile. In their own words:

> They must at all costs maintain their ignorance of a loved one's secret; hence the semblance of unawareness (nescience) concerning it. At the same time they must eliminate the state of secrecy: hence the reconstruction of the secret in the form of unconscious knowledge. This two-fold movement is manifest in symptoms and gives rise to gratuitous or uncalled for acts and words, creating eerie effects: hallucinations and delirium, showing and hiding that which, in the depths of the unconscious, dwells as the living-dead knowledge of *someone else's secret.* (1994, 188)

At this point, it is useful to bring to the fore Freud's distinction between mourning and melancholia because it provides the basis for my line of argumentation; by invoking Freud's dualistic conceptualisation I do not mean to imply that mourning is a healthy response to loss because it is all about letting the lost object go, while melancholia is the pathological condition of the subject which refuses to let the object go (Freud 1957). Rather, I would like to reexamine the mourning/melancholia dichotomy according to a queer ethics of grieving by which the refusal to let the object go—i.e. melancholia—becomes a political response to loss. David Eng and David Kazanjian, among others, have accepted Freud's distinction between mourning and melancholia and still argue that melancholia is preferable to mourning because "it is an enduring devotion on the part of the ego to the lost object, and as such is a way of keeping the other, and with it the past, alive in the present" (2003, 3).[5] If, as these authors suggest, to keep the object might imply a resistance to letting it go and therefore points to an ethical manoeuvre of "enabling, rather than blocking new forms of attachments" (Ahmed 2004, 159), then, there is a sense in which *The Night Watch*'s characters comply with this premise, oscillating between hopeless affects and the urge to render them visible, if only to dissipate their original semantic and ontological violence.

Other authors like J. E. Muñoz (1999) and Ann Cvetkovick (2003) propose a model of non-pathological melancholia that, following Raymond Williams, is defined as a "structure of feeling" (Williams 1975), a structure of everyday life that is not necessarily negative. This means that for queers—as well as for people of colour—melancholia is not a pathology but rather an identity-affirming tool of survival:

> It is a mechanism that helps us (re)construct identity and take our dead with us to the various battles we must wage in their names—and in our names [. . .] This melancholia is a productive space of hybridization that uniquely exists between a necessary militancy and indispensable mourning. (Muñoz 1999, 74)

This calling for militancy and activism through personal and/or collective traumas has been essential for queer people as a way of making visible an ethics of queer affects, which, otherwise, would have remained silenced. Not coincidentally, in this novel Sarah Waters describes a set of destructive affects pertaining to queer lives, rendering them visible and opening new paths of interpretation for the literary critic.

From an examination of the different embodiments of shame as an invisible affect that is attached to all and each of the protagonists of *The Night Watch*, there emerges a sense in which the epistemological and ontological scope of this narrative can be broadened. While recognising the potential of negative affects to be transformed into renewal and hope, I have also referred to the difficulties and psychic resistance that the four main characters find as integral parts of their existence. Remarkably enough, in her interesting study on happiness, Sara Ahmed (2010) argues that the fantasy of happiness is linked to a set of social indicators that function as predictors of happiness such as marriage, stable families and communities. Moreover, she criticises classical descriptions of happiness that promote a specific profile of happy people as typically white, heterosexual, married and with children: by considering happiness in that way, happiness becomes a privileged site for the establishment of social, moral and cultural hierarchies that are already given in advance as happiness indicators. Ahmed's task is not so much to dislodge happiness from certain objects as to make us aware of how happiness involves the display of optimistic affects towards certain objects. She bases her argument on the idea taken up by scholars such as Tomkins or Sedgwick, whereby affects are considered to be contagious. According to Ahmed, affects are sticky and depend on the objects and/or persons they are attached to, thus being dependent on a structural contingency in which the role of reception becomes essential. As she puts it: "to be affected by another does not mean that an affect simply passes or leaps from one body to another. The affect becomes an object only given the contingency of how we are affected. We might be affected differently by what gets passed around" (2010, 39).

Indeed, the four protagonists could have been affected differently by discarding painful feelings and sticking to joyful emotions instead—after all, all of them experience intense feelings of love and friendship. Yet, the prevalence of extreme frustration, blockage and numbness caused by both the individual traumas and the suffocating atmosphere of wartime may entail the novel's commitment to a particular way of telling as an ethical stance. In other words, the reverse chronology acts here as a creative vehicle to rescue all those unacknowledged experiences that lie at the core of the protagonists' traumas and silences since only by digging into the foundations of shameful and unbearable secrets can the reader unveil their cause and understand them. Thus, Kay's inert and phantom-like attitude, Helen's paranoia and self-destructive feelings, Viv's sadness and restlessness, and Duncan's utter shame and apathy cannot be understood unless their past is revealed. The chronological jump from 1947 to 1941 allows the showcasing of both the roots of the protagonists' negative affects and the struggle to envision their future hopefully. Such a formal strategy is inherently linked to the production of an alternative genealogy of queer lives, to borrow Halberstam's concept, in which configurations of new meanings can emerge.

In using this narrative structure, Waters' novel resorts to Benjamin's theses on discovering new ways of dealing with a past whose meaning lay "full phantom deep" but which, nevertheless, is subject to "selecting its precious fragments from the pile of debris" (Benjamin 1968, 258). In Benjamin's view, the figure of the collector is a key one when it comes to digging under the layers of language and concepts, for in moments of personal and collective rupture and displacement, the search for a creative rethinking of the past may fuel a more profound transformation in one's life. Similarly, in order to decipher and understand the melancholic shame that lies behind "that curtain coming down" (Waters 2006, 17) one must attend the anticlockwise narrative movement as a means of retelling the past and shedding light on the present.

As has been argued throughout this chapter, then, the resignification of shame and melancholia brings about a new model of affects that can emerge out of a self-conscious relation to the other, evoking an ethical commitment to defying pathological definitions of melancholic shame. Finally, I hope I have highlighted the importance of bringing the affect of shame into the theoretical and analytical discussion of trauma and literary studies and, accordingly, discussing it under different theoretical lenses, since to move beyond master narratives of trauma necessarily implies to put forward alternative and complementary models of interpretation and analysis. Moreover, the scholarly neglect of shame can be counteracted by narratives such as *The Night Watch* in which there exists the possibility of enacting an affective fullness, transcending teleological configurations of affects.

*The research carried out for the writing of this chapter is part of a project financed by the Spanish Ministry of Economy and Competitiveness (MINECO) (code FFI2012-32719). The author is also grateful for the support of the Government of Aragón and the European Social Fund (ESF) (code H05).

NOTES

1. When defining the identities of the protagonists as queer, I do not only mean in terms of their sexuality (Kay, Mickey, Helen and Julia are lesbians, and Duncan is a gay man) but also in that one character, Viv, does not stick to heteronormative practices of marriage and reproduction.

2. All quotations from this work have been taken from Sedgwick and Frank's 1995 compilation of texts from Tomkins' *Affect, Imagery, Consciousness*. 4 volumes. New York: Springer. For practical reasons, I have used Sedgwick and Frank's edition in my references to Tomkins.

3. These authors, informed by deconstruction and poststructuralist theories, have extensively contributed to highlighting a performative and antiessential-ist definition of shame. Moreover, they move beyond the human/non-human dualism of emotions so as to articulate an "ethics of flourishing" (Cuomo 1998) that challenges hierarchical definitions of affects in favour of a more equalitarian and respectful position towards animals, nature and inanimate objects. Only by acknowledging that certain types of human violence are necessarily bound up with the systematic mistreatment of non-human communities can we surpass univocal and hegemonic epistemologies.

4. Even though Julia is not a central protagonist, she is portrayed as one of the most positive characters: intelligent, kind, loving, and her manners always polite. Similarly, we also find Mickey, Kay's friend and workmate at the ambulance, as another example of goodness, generosity and care. Despite their class differences—Mickey is a working-class butch working as a mechanic in a garage after the war—both women personify rewarding affects. At the same time, though, we can trace a reverse parallelism between Kay and Helen, since the former's upper-class and the latter's working-class status do nothing but reunite them as the emblem of solitude, grief and loss. The categorisation of affects based on class differences is defied in this novel since economic comfort does not necessarily imply a happier life.

5. Such a statement echoes Walter Benjamin's unorthodox theses on history and philosophy as dialectical sites of fragmented materials and concepts one should bear witness to in the sense of pursuing a creative act of rethinking the past, just as a collector or an artist might do (Benjamin 1968, 59–67). It is in this Benjaminian spirit of remembering and embracing the deviant and the other that Eng and Kazanjian define melancholia as "an ongoing and open relationship with the past—bringing its ghosts and specters, its flaring and fleeting images, into the present" (2003, 4).

WORKS CITED

Abraham, Nicholas, and Maria Torok. 1994. *The Shell and the Kernel: Renewals of Psychoanalysis*. Edited, translated and introduced by Nicholas T. Rand. Chicago and London: The University of Chicago Press.

Ahmed, Sara. 2004. *The Cultural Politics of Emotion*. Edinburgh: Edinburgh University Press.

———. 2006. *Queer Phenomenology: Orientations, Objects, Others*. Durham and London: Duke University Press.

———. 2010. *The Promise of Happiness*. Durham and London: Duke University Press.

Benjamin, Walter. 1968. *Illuminations: Essays and Reflections*. New York: Schocken Books.

Braidotti, Rosi. 1993. *The Emotional Tie: Psychoanalysis, Mimesis and Affect.* Stanford: Stanford University Press.

Budden, Ashwin. 2009. "The Role of Shame in Post-Traumatic Stress Disorder: A Proposal for a Socio-Emotional Model for *DSM-V.*" *Social Science and Medicine* 69: 1032–39.

Butler, Judith. 1993. *Bodies that Matter: On the Discursive Limits of Sex.* London and New York: Routledge.

———. 1997. *The Psychic Life of Power: Theories in Subjection.* Stanford: Stanford University Press.

Castle, Terry. 1993. *The Apparitional Lesbian: Female Homosexuality and Modern Culture.* New York: Columbia University Press.

Cuomo, Chris. 1998. *Feminism and Ecological Communities: An Ethic of Flourishing.* London and New York: Routledge.

Cvetkovich, Ann. 2003. *An Archive of Feelings: Trauma, Sexuality, and Lesbian Public Cultures.* Durham and London: Duke University Press.

Darwin, Charles. (1872) 1999. *The Expression of Emotions in Man and Animals.* Introduced by Paul Ekman. London: Harper Collins and Fontana.

Eng, David, and David Kazanjian, eds. 2003. *Loss: The Politics of Mourning.* Berkeley: University of California Press.

Freud, Sigmund. 1957. "Mourning and Melancholia." In *The Standard Edition of the Complete Psychological Works of Sigmund Freud,* vol. 14. Edited and translated by James Strachey. London: The Hogarth Press.

Halberstam, Judith. 2005. *In a Queer Time and Place. Transgender Bodies, Subcultural Lives.* New York and London: New York University Press.

Lewis, Michael. 1992. *Shame. The Exposed Self.* New York: Free Press.

Llewellyn, Mark. 2004. "'Queer? I Should Say It Is Criminal!': Sara Waters' *Affinity.*" *Journal of Gender Studies* 13 (3): 203–14.

Muñoz, José Esteban. 1999. *Disidentifications: Queers of Color and the Performance of Politics.* Minneapolis: University of Minnesota Press.

Munt, Sally. 2008. *Queer Attachments: The Cultural Politics of Shame.* Hampshire: Ashgate.

Pajaczkowska, Claire, and Ivan Ward, eds. 2008. *Shame and Sexuality: Psychoanalysis and Visual Culture.* London and New York: Routledge.

Sedgwick, Eve Kosofsky. 2003. *Touching Feeling: Affect, Pedagogy, Performativity.* Durham and London: Duke University Press.

Sedgwick, Eve Kosofsky, and Adam Frank, eds. 1995. *Shame and Its Sisters: A Silvan Tomkins Reader.* Durham and London: Duke University Press.

Tomkins, Silvan S. 1962–1992. *Affect Imagery Consciousness.* New York: Springer.

Tomkins, Silvan. 1995a. "Shame-Humiliation and Contempt-Disgust". In Eve K. Sedgwick & Adam Frank. *Shame and Its Sisters: A Silvan Tomkins Reader.* Durham and London: Duke University Press, pp.133–178.

Tomkins, Silvan. 1995b. "What Are Affects?" In Eve K. Sedgwick and Adam Frank. *Shame and Its Sisters: A Silvan Tomkins Reader.* Durham and London: Duke University Press, pp. 33–74.

Waters, Sarah. 2006. *The Night Watch.* London: Virago.

15 "Welcome to contemporary trauma culture"

Foreshadowing, Sideshadowing and Trauma in Ian McEwan's *Saturday*[*]

Bárbara Arizti

"Groping back to bed after a piss" at four o'clock in the morning, the poetic persona of Philip Larkin's "Sad Steps" parts the "thick curtains" and is "startled by/ The rapid clouds, the moon's cleanliness" and the "wedge-shadowed gardens" (Larkin 2000, 2569).[1] The scene is for him a poignant "reminder of the strength and pain/ Of being young [that cannot] come again." Henry Perowne, the main character of McEwan's *Saturday* (2006), whose daughter's favourite poet is Larkin, wakes some hours before dawn and similarly moves towards the bedroom window. He does not immediately understand what he sees: a meteor burning out in the London sky? A comet? It is in fact an aeroplane in flames approaching Heathrow. Although the most plausible explanation seems to be mechanical failure, Perowne fears this is one more instance of Islamic terrorism. The nostalgic musings of Larkin's poetic persona are instantly overshadowed by Perowne's concerns, which refer the reader to the arena of current international affairs. Set in London, some eighteen months after the 9/11 attacks and only a month before the invasion of Iraq, *Saturday* focuses on the way large-scale terrorism has altered contemporary experience.

"Welcome to contemporary trauma culture," greets Roger Luckhurst (2008, 2) in his book *The Trauma Question*, a study of the proliferation of the idea of trauma from the year 1980, when the American Psychiatric Association officially added Post-Traumatic Stress Disorder (PTSD) to its list of mental illnesses. "One can now read up on the traumas that drive post-war Germany, post 9/11 America, Eastern Europe after Communism, or post-colonial Britain," he adds (2). Quoting Cathy Caruth, he considers trauma "a symptom of history" that "extends beyond the bounds of a marginal pathology and has become a central characteristic of the survivor experience of our time" (4). The terrorist attacks on New York, which loom large in the plot of *Saturday*, are a clear example of what Luckhurst calls the "iconic trauma events" (1) that have convulsed contemporary Western cultures. Similarly, Paul Crosthwaite (2009, 9) includes the collapse of the World Trade Center in his list of "seismic moments," events, scholars agree, that have made a major contribution to the postmodern *zeitgeist*, which, Crosthwaite argues, has reformulated history in terms of trauma.

The postmodern unfolds between the evocation of traumas past, mainly the Second World War, and the anticipation of a disaster yet-to-come, "the onset of nuclear annihilation," Crosthwaite states (43), or the fear of a terrorist attack, one could add. McEwan's novel is pervaded by the spirit of "dreading forward" (Stonebridge, quoted in Crosthwaite, 43), the fear that terror could strike again and affect one or one's people. As Jean-Michel Ganteau puts it, "trauma has become both immanent and imminent" (2011, 30). The Emergency Plan at the hospital where Perowne, a neurosurgeon, works has been revised to include "words like 'catastrophe' and 'mass fatalities,' 'chemical and biological warfare' and 'major attack',," words, Perowne says, that "have recently become bland through repetition" (10).

But trauma in *Saturday* does not exclusively come from historical sources. The traumas inherent to the human condition like aging or physical and mental illness are also a major concern, as are the personal traumas of Henry Perowne, whose Saturday routine is shattered by an unexpected event. The novel approaches the concept of trauma as a "nodal point," a "switching centre" (Luckhurst 2010, 203) where specialist and general uses of the term blend and clash. Trauma as a collective condition of our times, and as a disturbing individual experience coming from unforeseeable circumstances, and, to lesser extent, trauma as a clinical state, take prominence over the others. As Laura Marcus states, McEwan's fiction is "preoccupied by the relationship between public and private histories, and with the ways in which our individual experiences occur in tandem, or at odds, with historical events" (2009, 84). This paper intends to explore the interdependence of private and public trauma in the novel both on a thematic and on an aesthetic level. I will resort to the ideas of Gary Saul Morson on the representation of narrative time and to Elena Semino's theory of possible narrative worlds, in order to explore the paradoxical ways in which the novel both encourages and resists determinism in the face of traumatic events. It is my contention that the clash between an open and a closed form of temporality that pervades the novel is resolved differently in the public and private arenas. While historical events are perceived through the lens of "backward causation," the handling of fictional temporality in the private sphere promotes responsibility, agency and freedom of choice. The figure of the protagonist will be analysed in this respect as an example of how to go on living meaningfully in the midst of personal and historical violence. The role played by intertextuality will also be explored, as the process of overcoming anxiety is presented in the form of literary "repetition compulsion."

"The present," states Mark Currie (2010, 8), "as philosophy knows well, doesn't exist, and yet it is the only thing which exists. The past has been, and so is not, and the future is to be, and so is not yet. That only leaves the present." The main character of *Saturday*, a one-day novel written in the present tense, is very much aware of time. In *About Time*, a study on narrative, fiction and the philosophy of time, Currie devotes some pages to McEwan's novel as an example of the interaction between fictional and philosophical temporality. He draws attention to the novel's connections with *Mrs Dalloway* and *Ulysses* and their concern with internal and external time, the time of the mind and the time of the clock (129): "Like *Mrs Dalloway*, this is a novel which shows a constant interest in the linearity of clock time, and through constant reference to clocks, the reader always knows the time to within a few minutes" (130). Perowne's thoughts, available to the reader in free indirect style, often revolve around the present, "understood as the contemporary world" (Currie 2010, 9): "It is in fact the state of the world that troubles him most" (McEwan 2006, 80). International affairs penetrate the novel in the form of the TV news, to which Perowne has developed a kind of addiction: "a condition of the times, [. . .] to hear how it stands with the world, and be joined to [. . .] a community of anxiety" (180).

Perowne, however, does not only experience international affairs vicariously. The news bulletins act as a refrain throughout the novel, featuring, among other things, the story of the plane on fire that Perowne witnessed in the early hours. The concept of "monumental time," which Ricoeur coined in his analysis of *Mrs Dalloway* to refer to the "complex apparatus of public history, collective experience and authority that constitutes the backdrop against which the private thoughts and actions of the characters are staged" (Currie 2010, 129), also applies to McEwan's novel. According to Currie, the incident of the burning plane at the beginning of Perowne's day performs an important role in transforming clock time into monumental time: "it is through the rolling reports of TV news that this incident passes from the realm of a private occurrence into the public domain" (130).

February 15 2003, the day on which the novel is set, is marked out by the London peace march against the invasion of Iraq, also featuring in the TV news, a fact that helps situate Perowne in his historical moment. The TV news, Currie states, is a kind of clock by which the protagonist measures his private experiences, "but it is also the unfolding story of the historical day, through which the contemporary historical context of the day [. . . finds its way] into the novel" (2010, 130). Although the protagonist does not take part in the demonstration, he encounters the marchers on his way to his squash game with a colleague, part of his Saturday routine. He regards them with a combination of scepticism and admiration. The main reason why he does not join in the protest is that he has mixed feelings about the coming invasion. Saddam's police tortured one of his patients, an Iraqi professor. Henry saw his scars and listened to his stories (McEwan 2006, 60). This fact prompts a reflection on the accidental nature of opinions: "if he hadn't met and admired the professor, he might have thought differently, less ambivalently, about the coming war. Opinions are a roll of the dice" (72). In her contribution to the collective volume *Cognitive Poetics*, Elena Semino (2003, 83) analyses the different worlds that coexist in a narrative text. She makes a basic distinction between "what counts as 'reality' in the story"—Perowne's having met the Iraqi professor and his mixed feelings about the invasion—"and alternative, unrealised ways in which that reality might have turned out"—the fact that had he not met the professor he might have thought less ambivalently about the war. The opposition between the "text actual world' and 'textual alternative possible worlds" (Semino 2003, 86) is brought to the fore in *Saturday* through the reflections of the protagonist. The issue is directly thematised when Perowne, gazing at the plane in flames, remembers a famous thought experiment known as Shrödinger's cat, in which a cat in a closed container can be either still alive or might just have been killed by a random mechanism. "Until the observer lifts the cover from the box," remembers Perowne, "both possibilities, alive cat and dead cat, exist side by side, in parallel universes, equally real" (McEwan 2006, 18). Later in the day, when he learns from a news bulletin that the crew escaped unscathed, he comments: "Schröedinger's dead cat is alive after all" (36).

In *Narrative and Freedom: the Shadows of Time* (1994), Gary Saul Morson studies the different ways time is represented in fiction and their implications for how we live and think about our lives. Perowne's interest in possibility and in parallel universes co-existing side by side, can be read, in Morson's terms, as an example of "sideshadowing," an open fictional temporality, which "projects—from the 'side'—the shadow of an alternative present" (11). Sideshadowing signals a middle realm of possibilities that could have happened even if they did not. As Morson puts it: "Things could have been different from the way they were, there were real alternatives to the present we know, and the future admits of various paths" (69). He associates this form of temporality with human freedom, creativity, responsibility and contingency. The opposite of sideshadowing is "foreshadowing," a closed form of temporality that encourages determinism:

> When foreshadowing is used, certain events take place in a special way. Instead of being caused by prior events, they happen (or also happen) as a consequence of events to come. Foreshadowing, in short, involves backward causation, which means that, in one way or another, the future must already be there, must somehow already exist substantially enough to send signs backward. Thus, if a writer should believe in fatalism, foreshadowing is an ideal way to convey this sense of time. (7)

The story of Schröedinger's cat is not the only instance of sideshadowing in *Saturday*. Significantly enough, at a certain point in the novel, Perowne speaks against determinism: "he never believed in fate or providence [. . .]. Instead, at every instant, a trillion trillion possible futures" (McEwan 2006, 128–29). Perowne still finds himself yearning for the unpredictable despite the fact that as he grows old he feels his chances are narrowing (28). In fact, in his self-analysis—Perowne is a very reflective character—he is aware of a certain pessimism. The story of the plane is brought up again: "he'd already voted for the dead" instead of simply considering the event "an accident in the making" (40).

The coming invasion of Iraq is the subject of many a conversation in the course of the novel. The war triggers off a series of predictions that range from those made by Henry's American colleague, who believes that Iraq will be liberated and democratised (McEwan 2006, 101), to those made by his two children, especially his daughter Daisy, for whom the invasion is utterly wrong and who thinks that "terrible things are going to happen" (191). Like the other characters, Perowne is sure that the war is inescapable (60). He is also sure that there will be revenge attacks on London and other Western cities. What he is uncertain about is the outcome of the war. He hopes for a short and effective incursion but at times fears that the occupation will be a mess (72). What if Blair was lying and there were no weapons of mass destruction? (143). In his argument with Daisy on the subject, however, the emphasis lies on an optimistic vision of the Iraq war: "it could be [. . .] the beginning of something better" (192).

Semino has investigated how texts construct intricate sets of states of affairs, which establish different ontological relations with each other. The processing of the text on the part of the reader involves, she says, "the incremental construction of networks of interconnected mental spaces," defined as "short-term cognitive representations of states of affairs, constructed on the basis of the textual input on the one hand, and the comprehender's background knowledge on the other" (2003, 89). Although in the diegesis of the novel all guesses about the future exist as side events, promoting indeterminism and openness, the author and the reader's background knowledge collapses the field of possibility. In 2005, when the novel was published, it was sufficiently clear that the worst-case scenario pictured by Daisy had come true. Bush was wrong (Bassets 2011, 6). His "freedom agenda" had failed. His war against Sadam did not help spread democracy but utter chaos. His Global War on Terror brought about more terror. It was in 2004 that the Abu Ghraib prison, mentioned in the novel as Saddam's torture centre, became news for the acts of torture and abuse there committed by the U.S. Army. The terrorist attacks on Madrid and London are also a sad reality. Therefore, the fixing of narrative events retrospectively encourages a closed sense of temporality and tips the scales in favour of *foreshadowing*. There is a potent mise-en-abyme scene in the last pages of the novel that seems to confirm this view. Back at his bedroom window after an eventful day, Henry imagines a middle-aged doctor standing at this same window one hundred years before:

> February 1903. You might envy this Edwardian gent all he didn't yet know. If he had young boys, he could lose them within a dozen years, at the Somme. And what was their body count, Hitler, Stalin, Mao? Fifty million, a hundred? If you described the hell that lay ahead, if you warned him, the good doctor—an affable product of prosperity and decades of peace—would not believe you. (McEwan 2006, 286)

This affable Edwardian gent and Henry Perowne mirror each other in more than one sense. Perhaps, the most noticeable difference between them is the fact that Perowne would not find it so difficult to believe in the traumatic events that the future holds in store. As Freud argued, the traumas of the future are built upon those of the past: "anxiety—expectation of a *future* danger—itself possesses the potential to summon up painful memories of the past" (Crosthwaite 2009, 43).

Driving through London on his way to his squash game, Perowne thinks of the threat terrorism poses to him and his family (McEwan 2006, 80). Paradoxically, in the end, the threat does not come from international affairs but from his irresponsibility in handling a private event. His world is endangered not by Muslim terrorists but by his own blindness (Wall 2008, 762). It is to the personal traumas of Perowne's *Saturday* that I now turn, an arena on which the clash between foreshadowing and sideshadowing, determinism and freedom, will be finally resolved in favour of the latter.

On his drive through London, the protagonist causes minor damage to a red series-five BMW, a vehicle, Perowne admits, he "associates for no good reason with criminality" (McEwan 2006, 83). His suspicions are confirmed and an incident with the car's three occupants follows. Perowne refuses to apologise and standoffishly rejects the cigarette the other driver offers. When he is on the point of being beaten, he ventures a quick diagnosis of the neurodegenerative disease he believes Baxter, the leader, suffers from: "'Your father had it. Now you've got it too.'" He feels like a "witch doctor delivering a curse" (95), the narrator adds. Perowne is right and this causes confusion in Baxter and his two mates, who had not been told of their leader's condition. Baxter feels humiliated and Perowne manages to escape unscathed. The whole scene is presented in fatalistic terms, encouraging the idea of backward causation. Even before the confrontation, Perowne feels "he's cast in a role, and there's no way out. This [. . .] is urban drama. A century of movies and half a century of television have rendered the matter insincere" (86); "everything, as soon as it happens, will seem to fit" (87). As Kathleen Wall (2008, 779) puts it: "the media have made this encounter into a parody of itself. It has taken away the particularity, the individuality of the actors." Huntington's disease provides the clearest example of foreshadowing in *Saturday*. Baxter is doomed by his genes. His future is "fixed and easily foretold": "between ten and twenty years to complete the course, from the first small alterations of character" to "dementia, total loss of muscular control, rigidity sometimes, nightmarish hallucinations and a meaningless end" (McEwan 2006, 94).

Chapter 3 offers a short respite. It ends on an epiphanic mood, as Henry is transfixed at listening to a song by his son Theo, a blues musician. "*Baby, you can choose despair, Or you can be happy if you dare*" (McEwan 2006, 175), he sings. Musicians, thinks Perowne, sometimes "give us a glimpse of [. . .] our best selves, and of an impossible world in which you give everything you have to others, but lose nothing of yourself" (176). The lyrics of Theo's song—opening up once more the field of possibility—and the allusion to self replenishment in the form of win-win game theory are only slightly diminished by Henry's use of the word "impossible."

The climax of the novel is reached in Chapter 4. Baxter and one of his mates, who had been following Perowne, unexpectedly appear at his door, holding a knife to his wife's neck: "*Of course*. It makes sense. Nearly all the elements of his day are assembled" (McEwan 2006, 213), Perowne thinks. Once in, they break his father-in-law's jaw, make his daughter undress and threaten to rape her and to kill Rosalind, his wife. Baxter knows he has no future and is therefore not afraid of consequences (217). Resolution comes in the form of a poem by Arnold Perowne's daughter is made to recite. This provokes a mood swing in Baxter and Theo is able to catch him unawares and push him off the stairs. The police come and Baxter is taken to hospital with a broken skull. Family tragedy has been avoided. Baxter's assault has been aborted. The family, all in various states of shock, gather around the table for dinner: "From across the kitchen comes loud, unnatural laughter" (242). Rosalind is the one to show the clearest symptoms of trauma in a novel, Ganteau states, that does not "elaborate on clinical evocations of individual trauma" but rather approaches it as "a strong emotional shock," in the ordinary sense of the word. Besides, adds Ganteau,

> the shock of the shattered family reunion, complete with the eruption of physical and psychological violence within the family cocoon, cannot be envisaged as traumatic yet, and will stand very little chance of degenerating into trauma as the last chapter takes care to evoke the collective de-briefing of the family while the protagonist is reminiscing, Molly-wise, at the end of the day, on the events of the previous hours. (2011, 30)

It is to the fifth and last chapter of the novel that I now turn, a chapter in which the character of Perowne will rise to his full stature and the scales will be tipped back towards sideshadowing.

According to Semino (2003, 87), alternative worlds can take different forms in a text. She mentions, among others, "intention worlds," which correspond to the characters' plans, and "wish worlds," which have to do with their wishes and desires. As Perowne knows, Baxter's assault on his family is not a product of fate but of his irresponsibility. He wishes he had handled things better that morning (McEwan 2006, 246). "I have to see this through. I'm responsible" (245), he tells his wife after a phone call from the hospital asking him to operate on Baxter. Rosalind's fear that he is planning revenge opens up again the field of possibility, but this time he will choose right. There is a clue to this just after Baxter's head hits the floor. "He could have left him die of hypoxia, pleading incapacity through shock," we are told (239), but instead Perowne opens his airway and places him in a recovery position. During the operation Perowne loses all sense of time and experiences the "pure present, free of the weight of the past or any anxieties about the future. [. . .] he's happier than at any other point on his day off, his valuable Saturday" (266). Differences notwithstanding, Dominick LaCapra's words on the overcoming of clinical trauma are relevant here: "To the extent one works through trauma [. . .] one is able to distinguish between past and present and to recall in memory that something happened to one (or one's people) back then while realizing that one is living here and now with openings to the future" (2001, 22). "We could have been killed and we're alive" (McEwan 2006, 280), Henry tells Rosalind, and at the end of the day, the future appears to him to be a "horizon indistinct with possibilities" (286).

Bach's Goldberg "Variations," which Perowne plays while operating, are a powerful symbol of the novel's new mood. After hesitating before the four recordings of the piece he owns, he finally makes up his mind in favour of Angela Hewitt's, containing all the repeats (McEwan 2006, 257). Later on, Perowne reflects on the circular design of Bach's masterpiece: how the final Aria, although apparently "identical on the page," is utterly transformed by all the variations that come before (262). During the operation, he wishes he could give the healing touch to Baxter's deteriorating brain (263), and afterwards, sitting by Baxter's bed, he feels for his pulse, quite an unnecessary gesture, since the monitor shows it is perfectly regular (271). He does it because he knows contact is reassuring for patients.

Like the Goldberg Variations, McEwan's book ends where it began, with Perowne in his bedroom. As happens with the Aria, he is changed by the events of this unusual Saturday. At this stage, the bookish character of the text becomes even more evident. The heavy intertextuality of *Saturday* has been commented on by several critics. Sebastian Groes distinguishes among three distinct types of intertextual engagement in the novel: "first, direct citation and the borrowing of 'voice'; second, the construction of parallels; and, third, echo and allusion." In the words of Jean-Michel Ganteau, *Saturday* is "a text that is very much conscious of belonging to the Modernist paradigm that McEwan has repeatedly revisited in his most recent novels" (2011, 30). *Ulysses*, "The Dead," and *Mrs Dalloway* are the most prominent intertexts. The knowing reader will derive extra gratification by spotting the references to these three masterpieces of Modernism. However, a comment by Laura Marcus (2009, 85) on the "split between the text's 'knowledge' that it is a repetition of a fictional predecessor" and the characters' "unawareness of this fact," seems to present intertextuality in a rather disturbing light. The characters, Marcus says, are "caught in the literary webs of echo, allusion, reference" (85). The artificiality deriving from the overwhelming reference to well-known intertexts certainly detracts from the idea of freedom and creativity encouraged by the practice of sideshadowing. Henry Perowne seems bound to act out the lives of his literary predecessors.

Critics do not seem to agree on whether Henry Perowne has evolved towards greater insight by the end of the novel. For some, he is still "the voice of white, male, professional-class privilege, deploying what Elaine Hadley calls a 'surgical act of liberal detachment' from other people, or rather from *the* people" (Clark Hillard 2008, 186). Others, however, believe that Perowne has gained in empathy and compassion: "Perowne now accepts a responsibility to (for) the other" (Knapp 2007, 141). For Caroline Bennett, his operating on Baxter "attains a salvific coloration: he saves and is saved by his actions, becoming redeemer and redeemed" (2008, 228). His becoming ego-less in the operating theatre (227) eventually frees him from his initial solipsism (225). In the opinion of Richard Brown (2008, 87), Perowne appears as "ethically mature" in the last stages of the novel. Many a critic evokes, directly or indirectly, the term "epiphany" to account for his transformation. McEwan draws on Joyce's "The Dead" and *Ulysses* for the portrait of the new Perowne. He watches Rosalind in her sleep with fondness, in the same way as Gabriel watches Gretta in Joyce's short story, and McEwan uses the word "haggard," a key word in "The Dead," to describe Perowne's reflection in the mirror that triggers off a meditation on aging. In bed, his thoughts evoke those of Molly in *Ulysses*: "He feels his body, the size of a continent, [. . .] he's a king, he's vast, accommodating, immune, he'll say yes to any plan that has kindness and warmth at its heart" (McEwan 2006, 279). Larkin's "Sad Steps" is invoked once more when Perowne gets up to relieve himself in the middle of the night and stops at the window. The sacramental vision critics have identified in Larkin's poetry also has a bearing here. His poetic personas, agnostic though they are, yearn for a spiritual vision that seems out of reach (King 2012; Swarbrick 1995). Perowne, an inveterate materialist and a literary philistine who remains unmoved by the novels his daughter recommends, comes close to an epiphanic experience and finally acquires a sense of community. And who better than Gabriel Conroy to look for inspiration?

> In "The Dead" Gabriel, after becoming *aware* of his limitations as a lover and public figure, lapses out of his paralytic self-consciousness to become part of the reality of being. In a sense, his epiphany is a necessary prelude to his suspension of ego and will. But the reader is aware that Joyce is performing the suspension of consciousness and moving beyond Gabriel's awareness to imply the value of human love, particularly between a man and a woman. (Schwarz 1994, 67)

Joyce's character is looking out of the window as well when his ego dissolves and his vision expands to encompass his wife, his elderly aunts, his wife's ex-boyfriend long dead, the whole of Ireland and, finally, all humanity through the snow falling faintly upon all the living and the dead, an ending echoed in *Saturday*'s closing words: "And at last, faintly, falling: this day's over" (McEwan 2006, 289). Still at the window, Henry reflects on his day and anticipates the future. In the same way as Gabriel muses over his aunt's impending death, Perowne imagines his mother's and his father-in-law's death as well. "What else, beyond the dying?" he wonders. On a more positive note, he pictures Theo making his first move from home and Daisy publishing her poems and giving birth to her baby (285).

There are other things he knows must happen. Inexorable, Baxter's illness will follow its course. Perowne makes a mental note to persuade his family and the police not to press charges: "By saving his life in the operating theatre, Henry also committed Baxter to his torture. Revenge enough" (McEwan 2006, 288). Large-scale trauma, which has served as a backcloth to the family's ordeal, reappears at the end of the day: The war will start next month (287) and a terrorist attack on London is inevitable (286). The reader in the know will flinch at the thought. The only consolation might come from the oft-quoted words of Spinoza, the Dutch Jewish philosopher, for whom peace was not just the absence of war, but a virtue, a state of mind, a disposition for benevolence, which is precisely what Henry Perowne has acquired in the course of his eventful Saturday. The analysis of the different forms of trauma and fictional temporality in McEwan's novel proves that it is in the realm of the private that tensions are finally worked out. Individual agency, private responsibility, and, above all, professional deontology, seem to be McEwan's formula for the transformation of the world stage.

*The research carried out for the writing of this chapter is part of a project financed by the Spanish Ministry of Economy and Competitiveness (MINECO) (code FFI2012–32719). The author is also grateful for the support of the Government of Aragón and the European Social Fund (ESF) (code H05).

NOTES

1. Excerpts from "Sad Steps" from *The Complete Poems of Philip Larkin* by Philip Larkin, edited by Archie Burnett, Copyright © 2012 by The Estate of Philip Larkin. Reprinted by permission of Farrar, Straus and Giroux, LLC, and Faber and Faber Ltd.

WORKS CITED

Bassets, Lluís. 2011. "George Bush no tenía razón." *El País*, March 10.
Bennet, Caroline. 2008. "'Pathological Rationality' in Ian McEwan's *Enduring Love* and *Saturday*." *American, British and Canadian Studies* 10: 222–35.
Brown, Richard. 2008. "Politics, the Domestic and the Uncanny Effects of the Everyday in Ian McEwan's *Saturday*." *Critical Survey* 20 (1): 80–93.
Clark Hillard, Molly. 2008. "'When Desert Armies Stand Ready to Fight': Re-Reading McEwan's *Saturday* and Arnold's 'Dover Beach'." *Partial Answers* 6 (1): 181–206.
Crosthwaite, Paul. 2009. *Trauma, Postmodernism, and the Aftermath of World War II*. New York: Palgrave MacMillan.

Currie, Mark. (2007) 2010. *About Time: Narrative, Fiction and the Philosophy of Time*. Edinburgh: Edinburgh University Press.

Ganteau, Jean-Michel. 2011. "Disquieted Negative Capability: The Ethics of Trauma in Contemporary Literature." In *Between the Urge to Know and the Need to Deny: Trauma and Ethics in Contemporary British and American Literature*. Edited by Dolores Herrero and Sonia Baelo-Allué, 21–36. Heidelberg: Universitätsverlag Winter.

Groes, Sebastian. 2009. "Ian McEwan and the Modernist Consciousness of the City in *Saturday*." In *Ian McEwan. Contemporary Critical Perspectives*. Edited and introduced by Sebastian Groes, 99–114. London: Continuum.

King, Don W. 2012. "Sacramentalism in the Poetry of Philip Larkin." Accessed March 25. http://www.mrbauld.com/larkinv1.html.

Knapp, Peggy A. 2007. "Ian McEwan's *Saturday* and the Aesthetics of Prose." *Novel: A Novel on Fiction* 41: 121–43.

LaCapra, Dominick. 2001. *Writing History, Writing Trauma*. Baltimore & London: The Johns Hopkins University Press.

Larkin, Philip. (1962) 2000. "Sad Steps." In *The Norton Anthology of English Literature*. Edited by M. H. Abrahams and Stephen Greenblatt. Vol. 2. New York and London: Norton and Company.

Luckhurst, Roger. 2008. *The Trauma Question*. Abingdon: Routledge.

———. 2010. "The Trauma Knot." In *The Future of Memory*. Edited by Richard Crownshaw, Jane Kilby and Anthony Rowland, 191–206. New York and Oxford: Berghahn Books.

Marcus, Laura. 2009. "Ian McEwan's Modernist Time: *Atonement* and *Saturday*." In *Ian McEwan: Contemporary Critical Perspectives*, edited and introduced by Sebastian Groes, 83–98. London: Continuum.

McEwan, Ian. (2005) 2006. *Saturday*. New York: Anchor Books.

Morson, Gary Saul. 1994. *Narrative and Freedom: The Shadows of Time*. New Haven and London: Yale University Press.

Schwarz, Daniel R., ed. 1994. *The Dead*. Boston and Ithaca, New York: Cornell University Press.

Semino, Elena. 2003. "Possible Worlds and Mental Spaces in Hemingway's 'A Very Short Story'." In *Cognitive Poetics in Practice*. Edited by Joanna Gavins and Gerard Steen, 83–98. London and New York: Routledge.

Swarbrick, Andrew. 1995. *Out of Reach: The Poetry of Philip Larkin*. Basingstoke and London: Macmillan.

Wall, Kathleen. 2008. "Ethics, Knowledge and the Need for Beauty: Zadie Smith's *On Beauty* and Ian McEwan's *Saturday*." *University of Toronto Quarterly* 77 (2): 757–88.

Editors and Contributors

EDITORS

Marita Nadal is Professor of American Literature at the University of Zaragoza. She has coedited *Miscelánea: A Journal of English and American Studies* and the book *Margins in British and American Literature, Film and Culture*. Currently, her main fields of research are Gothic fiction and its relationship with trauma, and modern and contemporary US literature. Recent essays and publications include articles on William Golding, Edgar A. Poe, Joyce C. Oates, Flannery O'Connor and Shirley Jackson.

Mónica Calvo is Lecturer in American Literature at the University of Zaragoza. Her latest publications include the monograph *Chaos and Madness: The Politics of Fiction in Stephen Marlowe's Historical Narratives* (Rodopi, 2011). She currently specializes on the representation of trauma in contemporary American narrative, with a special focus on the existential component of compulsive behaviours.

CONTRIBUTORS

Marc Amfreville is Professor of American Literature at Paris-Sorbonne University. He has written many articles on nineteenth-century authors, three book-length studies: *Charles Brockden Brown: la part du doute* (Paris: Belin, 2000); *Pierre or the Ambiguities: l'Ombre portée* (Paris: Ellipses, 2003) and a two-hundred page essay on trauma and its representations in North American Literature: *Ecrits en souffrance* (Paris: Michel Houdiard, 2009), and a few papers on contemporary authors. He edited and revised the translation of Melville's *Pierre or the Ambiguities* (Paris: Gallimard, Bibliothèque de la Pléiade, 20005), he co-edited and revised the French translation of W. Faulkner's *Snopes' Trilogy*. He also translated and edited Charles Brockden Brown's hitherto unpublished in

France *Ormond, or the Secret Sharer* (1799) and more recently co-edited the first volume of F.S Fitzgerald's *Complete Works* in the Bibliothèque de la Pléiade (*This Side of Paradise* and *Flappers and Philosophers*, Paris: Gallimard, 2012). He wrote the chapter "American Gothic" in *A Literary History of America* (Cambridge, MA, Harvard UP, 2009) and a volume-length *Histoire de la littérature américaine* (together with Antoine Cazé and Claire Fabre; Paris: PUF, 2010).

Bárbara Arizti is Senior Lecturer in English Literature at the University of Zaragoza (Spain). She wrote her doctoral thesis on the work of David Lodge. Her current research interests are postcolonial literature and criticism, with special emphasis on the representation of ethics and trauma in Australian and Caribbean fiction. She has published widely in specialised journals and collective volumes and is the author of *Textuality as Striptease: The Discourses of Intimacy in David Lodge's Changing Places and Small World* (Peter Lang, 2002) and has co-edited *On the Turn: The Ethics of Fiction in Contemporary Narrative in English* (Co-editor, Cambridge Scholars Publishing, 2007).

Gerd Bayer is a tenured faculty member and Privatdozent at the University of Erlangen, having previously taught at Canadian and American universities. He has written on green narrative form in John Fowles's fiction, on postcolonial literature and film, on forms of Holocaust representations, and on popular culture. Most recently he has completed a book on English Restoration paratextual poetics and is now in the early stages of a new project on David Mitchell's work, concentrating on the idea of postcolonial cosmopolitanism.

Cathy Caruth is Frank H. T. Rhodes Professor of Humane Letters Graduate Faculty Member in the Department of English at Cornell University. She is author of *Unclaimed Experience: Trauma, Narrative, and History* (1996) and *Literature in the Ashes of History*, The Johns Hopkins UP (2013) and editor of *Trauma: Explorations in Memory* (1995). She is currently working on a book of interviews and photographic portraits titled *Pathbreaking Memories: Conversations with Leaders in the Theory and Treatment of Trauma*.

Maite Escudero Alías is Senior Lecturer in the Department of English and German Philology at the University of Zaragoza (Spain). She belongs to the research team "Contemporary Narratives in English" currently working on ethics and trauma in contemporary fiction in English. She has published widely on feminism, queer theory, cultural studies, trauma studies and affect theory, in journals such as *Journal of Gender Studies, Journal of Lesbian Studies, Journal of Transatlantic Studies* and *Journal of Popular Culture*. She is also the author of *Long Live the King: A Genealogy of Performative Genders* (2009).

Isabel Fraile Murlanch lectures in English Literature and English as a Second Language at the Faculty of Arts in Zaragoza (Spain). Her PhD thesis dealt with post-colonial and Australian studies, especially as reflected in the novels of Janette Turner Hospital. Her published research includes work on classical film noir, Hollywood comedy, Oscar Wilde and Australian literature. She is currently working on issues of identity, trauma, resilience, and processes of forgiveness. Her latest publication is "'Because a Man Is Born in a Stable . . . ' The Riddle of Australian Identity in Janette Turner Hospital's *The Last Magician*", in *Mapping Identities and Identification Processes: Approaches from Cultural Studies*. Ed. Eduardo De Gregorio. Long Hanborough: Peter Lang AG, International Academic Publishers (forthcoming).

Jean-Michel Ganteau is Professor of Contemporary British Literature at the Université Paul Valéry Montpellier 3 (France). He is the editor of the journal *Études britanniques contemporaines*. He is the author of two monographs (on David Lodge and Peter Ackroyd). He is also the editor, with Susana Onega and Christine Reynier, of nine volumes of essays on modernist and contemporary British literature and arts addressing such issues as impersonality and emotion, autonomy and commitment, the ethics of alterity, trauma and ethics, trauma and romance, etc. He has published articles on contemporary British fiction, with a special interest in the ethics of affects and trauma (as manifest in such aesthetic resurgences and concretions as the baroque, kitsch, camp, melodrama, romance).

Dolores Herrero is Senior Lecturer in English Literature in the Department of English and German Philology at the University of Zaragoza, Spain. Her main interests are postcolonial literature and cinema, on which she has published extensively. She co-edited, together with Marita Nadal, the book *Margins in British and American Literature, Film and Culture* (1997); together with Sonia Baelo, the books *The Splintered Glass: Facets of Trauma in the Post-Colony and Beyond* (2011) and *Between the Urge to Known and the Need to Deny: Trauma and Ethics in Contemporary British and American Literature* (2011). She was also the editor of *Miscelánea: A Journal of English and American Studies* from 1998–2006.

Avril Horner is Emeritus Professor of English at Kingston University, London. Her research focuses on Gothic fiction and women's writing. Major publications include *Daphne du Maurier: Writing, Identity and the Gothic Imagination* (1998) and *Gothic and the Comic Turn* (2005)—both co-authored with Sue Zlosnik; and, with Janet Beer, *Edith Wharton: Sex, Satire and the Older Woman* (2011). She edited *European Gothic: A Spirited Exchange 1760–1960* (2002) and is currently editing, with Anne Rowe, *Living on Paper: The Letters of Iris Murdoch 1934–1995*, which will be published by Chatto & Windus in 2015.

Aitor Ibarrola-Armendáriz teaches courses in migrant fiction, ethnic relations and film adaptation at the University of Deusto, Bilbao. He has published articles on minority and immigrant narratives, the pedagogy of literature and cinema, and processes of cultural hybridisation. Currently, he is the Director of the MA Program in Migrations and Social Cohesion (MISOCO) and Head of the Modern Languages and Basque Studies Dept. in the Faculty of Social Sciences and Humanities at the UD. He has edited several volumes: *Fiction and Ethnicity in North America* (1995), *Entre dos mundos* (2004), *Migrations in a Global Context* (2007).

Bilyana Vanyova Kostova is a research fellow at the University of Zaragoza where she has also completed her BA and Master's Degree in English Philology. Currently, she is working on her doctoral thesis on the problematic representation of trauma in the novels of Jeffrey Eugenides. Her main research interests are contemporary US fiction, trauma and memory studies, and ethnic literatures. She has given several talks at national and international conferences in Spain, the United Kingdom, and the Czech Republic.

Roger Luckhurst teaches in the School of Arts, Birkbeck College, University of London. He is the author of several books, including *The Trauma Question* (2008).

María Jesús Martínez-Alfaro is Senior Lecturer in English at the University of Zaragoza (Spain). Her research focuses on trauma and memory in contemporary literature in English, with a special emphasis on Holocaust trauma and literary representations of the Holocaust. She has published on these and other subjects in essay collections and journals like *Twentieth-Century Literature*, *JNT: Journal of Narrative Theory*, *Symbolism*, *Journal of the Short Story in English (JSSE)* and *The AnaChronisT*. She is one of the members of a research team currently working on the rhetoric and politics of suffering in contemporary narratives in English.

Susana Onega is Professor of English Literature at the University of Zaragoza (Spain) and the Head of a competitive research team currently working on ethics and trauma in contemporary fiction (http://cne.literatureresearch.net/). She has written numerous articles and book chapters on contemporary British literature and narrative theory and is the author of *Análisis estructural, método narrativo y "sentido" de The Sound and the Fury de William Faulkner* (Pórtico, 1980), *Form and Meaning in the Novels of John Fowles* (UMI Research Press, 1989. Winner of the Enrique García Díez Research Award, 1990), *Peter Ackroyd: The Writer and his Work* (Northcote House & The British Council, 1998),

Metafiction and Myth in the Novels of Peter Ackroyd (Candem House, 1999), and *Jeanette Winterson* (Manchester UP, 2006. Shortlisted for The ESSE Book Award, 2008). She is the editor of *Estudios literarios ingleses II: Renacimiento y barroco* (Cátedra, 1986) and of *"Telling Histories": Narrativizing History / Historicizing Literature* (Rodopi, 1995, 2006). She has introduced, edited and translated into Spanish John Fowles' *The Collector* (Cátedra, 1999) and has co-edited with José Angel García Landa *Narratology: An Introduction* (Longman, 1996. "Introduction" translated into Turkish, 2002), with John A. Stotesbury *London in Literature: Visionary Mappings of the Metropolis* (Carl Winter, 2001), with Christian Gutleben *Refracting the Canon in Contemporary Literature and Film* (Rodopi, 2004), with Annette Gomis *George Orwell: A Centenary Celebration* (Carl Winter, 2005), and with Jean-Michel Ganteau *The Ethical Component in Experimental British Fiction since the 1960s* (Cambridge Scholars Publishing, 2007), *Ethics and Trauma in Contemporary Narrative in English* (Rodopi, 2011), and *Trauma and Romance in Contemporary British Fiction* (Routledge, 2013). She is also the author of monographic sections on "John Fowles in Focus" in *Anglistik* 13 (1)/Spring 2002: 45–107, "Intertextuality", in *Symbolism: An International Journal of Critical Aesthetics*, 5 (New York: AMS. Press, 2005: 3–314), and on "Structuralism and Narrative Poetics" (Literary Theory and Criticism: An Oxford Guide, 2006: 259–79).

Silvia Pellicer-Ortín is Lecturer in the Department of English and German Philology at the University of Zaragoza (Spain). She wrote her PhD thesis on the work of Eva Figes as a research fellow at the University of Zaragoza and is currently a member of its research team, "Contemporary Narrative in English." She has been a visiting scholar at the Universities of Cambridge and Reading and has delivered many papers related to her main fields of research, mostly Trauma and Holocaust Studies, British-Jewish writers, autobiography, and feminism. She is the author of several articles dealing with these issues, and she has recently worked with Dr. Sonya Andermahr on the co-edition of the volume *Trauma Narratives and Herstory* (Palgrave Macmillan, 2013) and of a special issue of the journal *Critical Engagements* on the work of Eva Figes (2013).

Index

An environmentally friendly book printed and bound in England by www.printondemand-worldwide.com

#0014 - 270614 - C0 - 229/152/14 [16] - CB